Environmental & Safety Auditing
Program Strategies for Legal, International, and Financial Issues

Environmental & Safety Auditing
Program Strategies for Legal, International, and Financial Issues

Unhee Kim
with John Falkenbury
EEI Corporation

LEWIS PUBLISHERS

Boca Raton   New York   London   Tokyo

Acquiring Editor:	Joel Stein
Project Editor:	Suzanne Lassandro
Marketing Manager:	Greg Daurelle
Direct Marketing Manager:	Arline Massey
Cover design:	Jonathan Pennell
PrePress:	Kevin Luong
Manufacturing:	Sheri Schwartz

Library of Congress Cataloging-in-Publication Data

Kim, Unhee.
    Environmental and safety auditing : program strategies for legal, international, and financial issues / Unhee Kim, with John Falkenbury.
       p. cm.
    Includes bibliographical references and index.
    ISBN 1-55670-246-1 (alk. paper)
        1. Industrial management--Environmental aspects. 2. Environmental auditing. 3. Industrial safety I. Falkenbury, John. II. Title.
HD30.255.K56 1996
658.4′085—dc   96-27230
                                                                                       CIP

This book contains information obtained from authentic and highly regarded sources. Reprinted material is quoted with permission, and sources are indicated. A wide variety of references are listed. Reasonable efforts have been made to publish reliable data and information, but the author and the publisher cannot assume responsibility for the validity of all materials or for the consequences of their use.

Neither this book nor any part may be reproduced or transmitted in any form or by any means, electronic or mechanical, including photocopying, microfilming, and recording, or by any information storage and retrieval system, without prior permission in writing from the publisher.

The consent of CRC Press does not extend to copying for general distribution, for promotion, for creating new works, or for resale. Specific permission must be obtained in writing from CRC Press for such copying.

Direct all inquiries to CRC Press, Inc., 2000 Corporate Blvd., N.W., Boca Raton, Florida 33431.

© 1997 by CRC Press, Inc.
Lewis Publishers is an imprint of CRC Press

No claim to original U.S. Government works
International Standard Book Number 1-55670-246-1
Library of Congress Card Number 96-27230
Printed in the United States of America 1 2 3 4 5 6 7 8 9 0
Printed on acid-free paper

CONTENTS

PREFACE

Chapter 1
AUDITING 1

Chapter 2
CONTROVERSIES OF AUDITING 5
Risk and Benefit 5
Voluntary Versus Mandatory Program 5
Self-Evaluation Privilege 6
The Auditing Climate 6

Chapter 3
THE BENEFITS AND RISKS OF AUDITING 9
Benefits of Auditing 9
Risks of Auditing 12

Chapter 4
FEDERAL AGENCY POSITIONS
AND PERSPECTIVES 15
Department of Justice 16
Environmental Protection Agency 18
Auditing and Enforcement at Federal Facilities 32

Chapter 5
PRIVATE SECTOR INITIATIVES AND POSITION 45
Industry Survey 45
International Standards Organization 47
British Standards Institute and the Environmental Auditors
    Registration Association 48

Chemical Manufacturers Association 49
International Chamber of Commerce and Global
    Environmental Management Initiative 50
Environmental Auditing Roundtable 56
Coalition of Environmentally Responsible Economies 56
Corporate Environmental Enforcement Council 59
American Petroleum Institute 60
Other Organizations 63
Conclusion 63

Chapter 6
LEGAL PERSPECTIVES ON ENVIRONMENTAL,
HEALTH, AND SAFETY AUDITING 65
Concerns About Disclosure and Privilege 65
Traditional Common-Law Privileges 66
Federal Government Response 69
The State Legislators Respond 70
Major Case Histories 71
Enforcement Statistics 72
Occupational Safety and Health Administration 74

Chapter 7
AUDIT PROGRAM ISSUES 83
Commentary on Getting a Program Started 83
Audit Program Issues 87

Chapter 8
AUDIT-RELEVANT REGULATIONS 103
Occupational Safety and Health Act 103
Clean Air Act 104
Clean Water Act 104
Oil Pollution Control Act 105
Comprehensive Environmental Response, Compensation,
    and Liability Act 105
Emergency Planning and Community Right-to-Know Act 105
Pollution Prevention Act 106
Resource Conservation and Recovery Act 106
Toxic Substances Control Act 106
Endangered Species Act 107
Other Acts 107

National Environmental Policy Act and Environmental
  Quality Improvement Act 108
Executive Orders 110

Chapter 9
INTERNATIONAL STANDARDS ORGANIZATION
14000 REQUIREMENTS FOR AUDITING 111
Requirements of an Audit Program 117
Auditor Qualifications 118
Training 119

Chapter 10
SECURITIES AND EXCHANGE COMMISSION 121
Environmental Advantages 121
Industry Practice and Required Disclosures 122
Securities and Exchange Commission Guidance 125
Conclusion 129

Chapter 11
INTERNATIONAL AUDITS 131
International Regulations and Resources 131
Comments on International Auditing 133

REFERENCES 137

APPENDICES

Appendix A
ENVIRONMENTAL PROTECTION AGENCY POLICY 139
ENVIRONMENTAL AUDITING POLICY STATEMENT 140
CLARIFICATION ON POLICIES RELATED
  TO ENVIRONMENTAL AUDITING 157
INCENTIVES FOR SELF-POLICING: DISCOVERY, DISCLOSURE,
  CORRECTION, AND PREVENTION OF VIOLATIONS 170

Appendix B
INTERNATIONAL STANDARDS ORGANIZATION
DRAFT STANDARDS 189
ISO/CD 14010.2 "GUIDELINES FOR ENVIRONMENTAL
  AUDITING — GENERAL PRINCIPLES OF
  ENVIRONMENTAL AUDITING" 191

ISO/CD 14011/1.2 "GUIDELINES FOR ENVIRONMENTAL
   AUDITING — AUDIT PROCEDURES — PART 1: AUDITING
   OF ENVIRONMENTAL MANAGEMENT SYSTEMS" 197
ISO/CD 14012.2 "GUIDELINES FOR ENVIRONMENTAL
   AUDITING — QUALIFICATION CRITERIA FOR
   ENVIRONMENTAL AUDITORS" 206

Appendix C
ORGANIZATIONAL STRUCTURE OF ISO 14000/TC 207:
ENVIRONMENTAL MANAGEMENT AND UNITED STATES
PARTICIPATION CHART 213

Appendix D
GLOSSARY 217
DEFINITIONS 219
ACRONYMS 221

INDEX 225

LIST OF TABLES

Summary of the Environmental Leadership Program
   Pilot Projects 24
Principles for Environmental Management
   and Audit Elements 53
Sample Scoring Sheet for Principle 4:
   Employee Education 57
Criteria Factors for Audit Frequency 91
Summary of Analysis for Root Causes at the
   Management System Level 97
Comparison of Environmental Management Systems 114
Securities and Exchange Commission
   Required Disclosure 124

PREFACE

This book was published for use by local, state, and federal government and private sector personnel in an effort to encourage environmental safety and health professionals, industrial hygienists, junior and senior level corporate financial advisors, environmental lawyers, operational managers of national and international corporations, senior management personnel accountable to shareholders, and others responsible for environmental, safety, and health issues to look beyond the limit of the technical responsibilities of their jobs toward the management strategies of the entire program under their responsibility and to help auditors be more aware of all the auditing issues. This book brings the environmental issues and the safety and health issues together in one volume to assist in the development of an overall comprehensive audit program for your organization.

The opinions expressed in this book are solely those of the authors and do not in any way represent the policies or official positions of the United States Environmental Protection Agency, the Securities and Exchange Commission, or any other department or agency of the United States government. I have researched information from many government and private organizations to present this oversight of the legal, international, and financial issues and program strategies of auditing. This is a comprehensive overview of auditing principles and issues and an instruction on the benefits and risks of and strategies for developing an audit program. It is not an instruction on how to perform an audit, as there are many books and courses of that nature on the market today. It is a step above that level, with a focus on strategies for weighing all the legal, international, financial, and other issues. The book incorporates the accumulation and interpretation of the data received from or about the United States Environmental Protection Agency, the United States Department of Justice, the Securities and Exchange Commission, the International Chamber of Commerce, the International Standards Organization, the European Community Eco-Management and Audit Scheme, the Global Environmental Management Initiative, and the most current articles of other experts as noted in the text.

While writing this book, I have sought out many renowned professionals in the field and have learned much. Please keep in mind that auditing is an evolving discipline, and this book will require revision every few years to remain current. I hope that many professionals, new and established, will benefit from the resources I have compiled over the past year. I am thankful to all the other authors and contributors who so willingly contributed to the making of this book. A profile of the authors who have contributed to this book and their contributions follows this preface.

I would like to express special thanks to Mr. John Falkenbury for his invaluable assistance during the compilation of this book. In addition to his authorship of Chapter 8, he provided invaluable support as the second author and the technical editor.

Unhee Kim, MPH, CIH
President, EEI Corporation
May 2, 1996

AUTHORS

Unhee Kim, CIH is the founder of EEI Corporation and the primary author of this book. She has a B.A. in Biochemistry from Harvard University and a master's degree in public health from the Yale School of Public Health and Epidemiology. She is a certified industrial hygienist with expertise in health risk assessment and has a working knowledge of environmental issues. She received training related to auditing from the state of Rhode Island programs and from industry at the Hoechst Celanese Corporation. It was while she was the director of the Environmental and Industrial Hygiene Department of Tracor Technology Resources, Inc. that she focused on auditing programs and program development. Her fascination with this subject has culminated in the publication of this book.

John F. Falkenbury is a senior associate consultant for EEI. He has 26 years' experience in the environmental compliance field. Prior to working with EEI, Mr. Falkenbury held positions with federal regulatory agencies, the Department of Defense, the Department of Energy, and industry, administering environmental management programs and writing National Environmental Policy Act documents. He has developed environmental programs and written program documents, including Pollution Prevention and Hazardous Waste Minimization Strategies. He has developed and delivered training seminars on the National Environmental Policy Act and other environmental programs. He is well versed in the environmental laws and regulations, executive orders, and Council on Environmental Quality memoranda and has first-hand experience in implementing these regulations. He authored Chapter 8, verified data, and organized the book while writing sections of other chapters.

CONTRIBUTORS

Richard J. Satterfield is an environmental engineer and senior staff member of EPA's Federal Facilities Enforcement Office. Among other duties, he is presently chairing two interagency work groups, the Civilian Federal Agency Task Force and the Federal Agency Environmental Audit Work Group. Before joining the EPA, Mr. Satterfield served as the director of the Environmental Compliance Division at the United States Department of Commerce, where he had oversight responsibilities for the department's compliance with the federal, state, and local environmental regulations, presidential executives orders, and administrative orders. Mr. Satterfield made prominent contributions to Chapter 4.

Ralph Rhodes is the director of the Health, Safety and Environmental Audit for Allied Signal, Inc. He has directed the corporation's Environmental Audit Program since its inception in 1978. Prior to joining Allied Signal in 1977, Mr. Rhodes held public service positions for over 21 years in state and federal regulatory agencies, including positions at the EPA and its predecessor agencies.

Concurrent with the development and growth of the Environmental Auditing Program at Allied Signal, Mr. Rhodes has been actively involved in the development of environmental auditing as a professional specialty in the United States and Europe. He was a founder of the Environmental Auditing Roundtable and is currently a member of its board of directors. Mr. Rhodes has led over 300 environmental audits at a wide variety of Allied Signal plants around the world. He has made invaluable contributions to Chapters 7 and 11.

Timothy A. Wilkins, Esq. is an attorney in the environmental section of Bracewell and Patterson, L.L.P., in Houston, Texas. Mr. Wilkins obtained his J.D. at Harvard Law School and a master of public policy degree, concentrating in environmental policy, from the John F. Kennedy School of Government. He was the former editor-in-chief of the Harvard Environmental Law Review. Mr. Wilkins made significant contributions to Chapter 6 in this publication.

Brian P. Riedel is the head of the EPA's Office of Planning and Policy Analysis, which serves the assistant administrator for Enforcement and Compliance Assurance. Mr. Riedel is coauthor of the EPA's interim and final environmental audit policies. He is cochair of the Quick Response Team responsible for making recommendations regarding interpretation and application of the policies. In addition, Mr. Riedel is the Office of Enforcement and Compliance Assurance's lead on enforcement and compliance matters relating to ISO 14001 standards for environmental management systems. He is a member of the United States Technical Advisory Group to ISO Technical Committee 207 on environmental management. Before joining EPA, Mr. Riedel practiced environmental law with the Washington, D.C., law firm of Newman and Holtzinger. He received his law degree from the University of Wisconsin and B.A. from the University of Michigan. Mr. Riedel provided invaluable assistance and contributions in writing Chapter 4.

ADDITIONAL CONTRIBUTORS

Mr. Tai-Ming Chang is the director for the EPA's Environmental Leadership Program. He is with the Office of Compliance within the Office of Enforcement and Compliance Assurance. Tai-Ming has been with the EPA since 1984 working in a variety of programs, including Superfund and wetlands enforcement. He has a master's degree in public administration from Harvard University. He provided the research material on the Environmental Leadership Program and its pilot projects discussed in Chapter 4.

Ms. Jean H. Creary, Esq. is a member of the Environmental Practice Group of Nixon, Hargrave, Devans and Doyl, L.L.P., in Rochester, New York. She chairs the group's preacquisition, compliance, and asset-based lending audit practice. She has coordinated and conducted environmental compliance and preacquisition audits throughout the United States and internationally. Ms. McCreary serves as the president of the Environmental Auditing Roundtable, and chairs the committee developing the United States position on general auditing principles in the United States Sub-TAG of TC-207 of International Standards Organization. She has written the articles on the International Standards Organization activities and on the ISO 14000 Environmental Management Systems. She has been appointed as a representative of auditor interests by the American National Standards Institute to serve on the 18-member EMS council that is developing the American National Accreditation Program to accredit registrars, certifiers of auditors, and providers of training programs for ISO 14000. Ms. McCreary graduated from the University of Rochester and obtained her Juris Doctor at the University of Florida College of Law. She provided the technical review of Chapter 9.

Mr. Ronald Lund is the Corporate Environmental Audit Manager of Dow Corning Corporation. Mr. Lund joined Dow Corning in 1970 after obtaining a chemical engineering degree from the University of Utah. Experience at Dow Corning has been split between the manufacturing and engineering function, and the science and technology function. He is responsible for internal auditing of all global facilities and for risk assessments of contract waste disposers and manufacturers.

He is an active member of the EAR where he chairs the Risk Assessment Work Group, the Regulatory Matrix Group, and serves on the membership committee. He is a principle environmental auditor registered by the Environmental Auditors Registration Association in the United Kingdom. Mr. Lund provided a technical critique of Chapters 9 and 11.

David G. Sarvadi, Esq. is an associate with the Washington, D.C., law firm of Keller and Heckman and is also a certified industrial hygienist. He holds degrees from Pennsylvania State University, the University of Pittsburgh Graduate School of Public Health, and George Mason University School of Law. Mr. Sarvadi has worked in a number of different corporate and consulting environments. He currently represents clients before various state and federal regulatory agencies, including the Occupational Safety and Health Administration. He also counsels trade associations and individual companies on health and safety matters and assists them in presenting their views before various agencies in the regulatory process. He currently serves as counsel to the National Coalition on Ergonomics.

As an industrial hygiene consultant, he has assisted a variety of clients, e.g., the United States Department of State and the Internal Revenue Service, in setting up comprehensive health and safety programs, auditing corporate and plant level programs, and evaluating occupational exposures to industrial chemicals and processes. Mr. Sarvadi provided input to the safety section of Chapter 6.

Other professionals who have provided input through various discussions and documents include **Ms. Barbara J. McGuinness,** manager of Environmental Audit Programs, of E. I. DuPont de Nemours; and **Mr. Ather Williams, Jr.,** Vice President, Safety and Industrial Hygiene Worldwide, of Johnson and Johnson.

1 AUDITING

The term "audit" is applied in many disciplines, most often associated with environmental audits and financial audits. In the energy industry audits are most often performed for safety and health concerns, particularly at nuclear facilities. However, an audit is more than just self-analysis; there are many facets that must be considered when developing an audit program, such as the implications of disclosure to federal agencies, the changing legal issues on the state and federal level, or the new federal policy on self-policing. The program involves company commitment of all employees, especially that of management to a more active role, to be in control of benefits and risks, and to be ready for potential state and federal agency involvement.

A Price Waterhouse survey indicated that industry's main reasons for auditing are as follows:

- To identify problems internally and correct them before discovery by agency inspections
- To improve the company's overall environmental program and make it proactive
- To ensure that management's control systems are functioning
- To decrease the company's operating and financial risks (Ref. 1)

Management must understand from the start that there must be a commitment to correct deficiencies found during the audit. Deficiencies found and reported to senior management, but not corrected, pose serious personal risks to management. In addition, presentations of the audit findings to management should also include critical issues associated with the company's operations and known program weaknesses. The basic purpose of audit programs, *to evaluate programs and communicate deficiencies to higher management*, makes site management understandably reserved on the subject. Upper management support is needed to ensure cooperation at the plant level and to ensure that any deficiencies that are found are promptly corrected.

The audit program itself should be reviewed for its effectiveness both in identifying root causes at the management system level and at verifying closure of previous findings during subsequent audits. Any findings observed during the audit should be noted and *verified*. All findings should be clarified and identified with a root cause at the management system level.

An audit is an opportunity for an incoming manager to identify and document the existing status of affairs. By identifying all activities and action items, a baseline is established on which to make improvements and against which to measure performance. Such an audit is instrumental in setting the program goals, objectives, and priorities on a schedule. The audit report provides justification for annual budget requests, identifies appropriate staffing needs, and identifies target management integration concerns of all departmental functions.

The use of an audit to improve a company's compliance and internal environment is a two-edged sword. The Securities and Exchange Commission and the United States Environmental Protection Agency (EPA) have issued specific environmental disclosure requests to make the details about environmental programs and liabilities a matter of public record. Enterprise owners and operators have been extremely concerned with the question of disclosure. Members of the regulated community do not wish to perform self-analyses if forced disclosure of the findings will provide the government and other opponents the basis for penalties and law suits. The regulated community's concern about disclosure of self-audits, the common law tools for protecting reports, and the government agencies' approach to audits are discussed within this book.

Any voluntary information being used as a club against a company is troubling, especially when it is in the public interest that the information be disclosed. Agencies could use audit reports as a road map to a facility's possible areas of noncompliance. A company audit is performed to help identify and respond to environmental concerns and thereby prevent enforcement exposure, and such an audit should not be used to make the government's case against the company. If the agency makes the results of the audit part of the public record, the company may then have problems with its corporate image affecting its community relations.

One avenue to improved public relations is the company's annual report. An annual report can be a powerful means of a company's overall communication to its investors and the general public. Failure to mention any company environmental concerns would not go over well with the environmentally educated public that exists today. This means of communication can be further enhanced by the publication of a green report.

The many aspects of an audit program, the legal positions of industry and of the state and federal governments, the benefits and risks of auditing, and many other attributes of audit programs are discussed in this text. There are many formal definitions of "environmental audit" that are consistent with the

AUTITING 3

essential elements of *systematic, documented, objective, and periodic* evaluation. The International Chamber of Commerce (ICC) uses the European Community Economic audit definition:

> A management tool comprising a systematic, documented, periodic, and objective evaluation of how well environmental organization, management and equipment are performing with the aim of helping safeguard the environment by (i) facilitating management control of environmental practices; (ii) assessing compliance with company policies, which would include meeting regulatory requirements. (Ref. 2)

The EPA definition of an audit as it appears in the 1986 auditing policy is

> Environmental auditing is a systematic, documented, periodic and objective review by regulated entities of facility operations and practices related to meeting environmental requirements. (Ref. 3)

The International Standards Organization definition of an audit is as follows:

> A systematic, documented verification process of objectively obtaining and evaluating evidence to determine whether specified environmental activities, events, conditions, management systems, or information about these matters, conform with audit criteria, and communicating the results of this process to the client.

Aside from the formal definitions, practical definitions used in industry can be summarized as follows:

- An audit is a realistic assessment of a corporation's overall environmental status and the potential liabilities with respect to regulations, corporate goals, and programs.
- It is a quality control tool to measure environmental performance and the effectiveness of the environmental management system by reviewing management support and responsibility delegations, written programs and policies, reports and documents, standard operating procedures, actual operations, and interviews with employees.
- It is a snapshot assessment of a program in effect, designed to identify major and/or repeated incidences, or trends, but may not catch all transient occurrences.

Throughout the book the terms "audit" and "environmental audit" are used interchangeably. In many situations the term "environmental, safety, and health audit" is used. The "environmental, safety, and health audit" encompasses (is an umbrella for) the "environmental audit." Therefore some of the "environmental audit" issues and strategies discussed can be applied to the broader "environmental, safety, and health" audit.

Environmental enforcement and safety enforcement have always been separate operations within the federal government with two different agencies administering the regulations. However, practice has demonstrated a need to integrate and coordinate the two disciplines. Safety concerns are also environmental concerns, as often has been shown when findings from *separate* environmental and safety audits result in the identification of the same root cause. This book brings together the environmental auditing issues and safety auditing issues by demonstrating where the issues intersect and where they run parallel. The critical issues of environmental auditing may also apply to safety auditing, and the resolution of the environmental issues should also resolve the safety issues. Better integration of these two disciplines will successfully lead to less redundancy and better management of environmental safety and health affairs.

2 CONTROVERSIES OF AUDITING

After more than a decade of auditing experience, the EPA, the United States Department of Justice (DOJ), and other non-governmental organizations are coming together with industry sectors to face some persistent controversial issues of disclosure and fairness that have been escalating. Before trying to establish an audit program or improve an existing program, one should understand and appreciate the implications of the three target issues under discussion by all interested parties.

RISK AND BENEFIT

An audit is a double-edged sword, providing valuable environmental performance and liability assessment data, useful for internal management decision support in priority setting and corrective/preventive actions. However, the potential for information disclosure opens greater risk of civil and criminal investigations from enforcement agencies and litigation from other third parties. If the entire audit report is open to the public, industry faces greater risk of citizens' suit and criminal enforcement action based on purported knowledge of corporate officers. Any protective measure from the EPA or DOJ for corrective actions does not necessarily carry over to citizens' suits or other third-party litigation.

VOLUNTARY VERSUS MANDATORY PROGRAM

Voluntary initiative for sustainable development programs[1] and beyond compliance programs are sprouting up, not only in the United States, but also internationally. With our global economy and heightened environmental awareness, the private sector believes that it has to be environmentally proactive to be competitive and that mandated audits are unnecessary and counterproductive.

---

[1] Sustainable development programs are a new initiative for the management of company resources, environmental resources, finances, regulatory compliance, public image, for extended company growth and development. These programs are an umbrella to beyond compliance based programs.

The specter of government involvement introduces concerns over the type of outside control and is considered by industrial advocates to be a poor allocation of resources. At the same time, the regulatory environment is changing in many ways with no clear indication of what will be the requirements. The Congress has set a "freeze" on many regulations since 1995, but the re-authorization of many environmental laws are incorporating required auditing elements for permit issuance and for full compliance. Furthermore, because of the expertise that exists within industry, the industry audits are considered to be more effective than government inspections. Industry audits have the flexibility to use different types of auditing approaches, such as management, compliance, or transactional audits. The industry sector is interested in maintaining the audit program as an unregulated entity but with statutory protection of audit reports. On the other hand, many enforcement agencies are not satisfied with the insufficient use of auditing and are opting for mandated auditing programs.

SELF-EVALUATION PRIVILEGE

Current regulations and policies do not afford protection from disclosure, and this has a chilling effect on the openness of the audit program. At this time there is no statutory protection of the self-evaluation. Protection of self-evaluation findings is subject to judicial discretion, and therefore nothing is guaranteed. As a consequence, attorney-client privilege has been the means by which audit findings have been protected. However, the attorney-client privilege has been challenged in previous cases, where it has been upheld in some and not in others. Until self-evaluation findings are statutorily protected, the limits of attorney-client privilege protection for audit reports need to be understood by the audit industry to obtain maximum protection. The legal issues are more than just the disclosure and attorney-client privilege implication of an audit program, which in itself is central to minimizing the inherent risk. This subject is addressed in Chapter 6 to emphasize the importance of this issue in an audit program.

THE AUDITING CLIMATE

This is an opportune time to try and resolve the persistent controversial issues of environmental auditing that have been simmering and escalating over the last decade. With a collaborative attitude, a true demonstration of the EPA's new philosophy and approach, the EPA has initiated a major undertaking to encourage voluntary self-assessment. The EPA has re-assessed its current practice and policy on auditing and has determined their best alternatives for future direction in the new 1995 policy (Ref. 4). The EPA's Environmental Leadership Group has also selected 12 pilot projects to look at model audit programs and environmental management systems. Concurrently, the International Standards Organization's Technical Committee 207 is developing standards for environmental management systems that have specifications and guidance for auditing programs.

Many private sector organizations are actively advocating unregulated audit programs with self-evaluation privileges and are demanding competitive fairness for companies taking positive steps to investigate and correct past problems. Simultaneously, they are developing guidelines for audit programs. Some example audit program guidelines may be obtained from the Global Environmental Management Initiative (GEMI), the Environmental Auditing Roundtable (EAR), the Chemical Manufacturers Association (CMA), and the World Bank Group. Other sources are mentioned within the book.

3 THE BENEFITS AND RISKS OF AUDITING

Often, all the benefits of an audit program are not recognized or understood. Through an audit program useful data are collected and appropriate information can be transmitted both horizontally and vertically in an organization. There can be extensive benefits from an audit program if the organization and the audit program are properly structured to maximize the use of the collected data. To develop an auditing program that maximizes the benefits, you must first understand what are all the benefits.

BENEFITS OF AUDITING

1. *International Standards Organization Registration* — The International Standards Organization (ISO) 14000[2] is the standard series for environmental management systems (EMS). ISO 14000 registration has prerequisites for auditing programs that must be met prior to registration. Companies with ISO 14000 registration receive recognition of their environmental programs, which is a mark of excellence often utilized for green marketing.[3] Auditing requirements for ISO 14000 registration are discussed in Chapter 9.
2. *Management Decision Support Data* — Risk ranking is fundamental to management decision support both at the senior management level and the site level. Audit results provide objective, useful data for more accurate prioritization of environmental risks

---

[2] ISO is an independent international standards organization for industry. They are best known for the ISO 9000 Quality Standards. A technical committee for environmental management systems is currently developing the ISO 14000 EMS standards.

[3] Green marketing is the selling of a company's environmental philosophy, policy, and programs. Often, companies publish an annual "Green Report" with their annual fiscal report that summarizes their environmental programs, progress, and accomplishments.

and liabilities. This will assist in allocating resources to achieve overall maximum benefit for the company.

3. *Liability Assurance* — Audit practice increases the ability to divest properties. Having a history of an audit program will show that past liabilities have been identified and managed and therefore provides greater assurance to prospective buyers.

4. *Measurement Tool* — By practical definition, auditing is a measurement tool for assessing performance. As a tool it is especially useful in determining the progress toward integration of business and environmental goals for sustainable development of the organization. The integration of business and environmental goals will enhance the public perception of the organization. Use of auditing will ensure progress toward the organization's goals and should result in new growth.

5. *Compliance Management* — Regulatory compliance has become more complicated due to the overlap of regulations. The overlaps occur among federal regulations and between the federal and state levels. There has been a trend for increasing penalties for noncompliance in regard to specific monitoring and reporting requirements. An audit program can be a tool to manage the complex regulatory overload of the 259 sets of EPA regulations under review, development, or revision and the 670 training requirements for environment, safety, and health issues. Public awareness and demand is high for environmental quality assurance. Regularly scheduled audits help to monitor the resolution of previous findings and help identify new areas where compliance will be necessary due to amended or new regulations. To respond to the public and the regulatory demands, an audit program is necessary for compliance management.

6. *Educational Process* — The audit is an educational process for all site personnel. Through a preaudit questionnaire, the opening meeting, the actual audit, and the closing meeting, site personnel will become aware of those areas under their purview that require attention. The preaudit questionnaire is a search process to identify areas of concern, previously unidentified and unknown risks, and a compilation of documentation of current practices. Senior management gains a great deal of information by participating in the opening meeting. They will learn what will be audited, the process by which the information will be gathered, the time table, the list of personnel to be interviewed, how the findings will be handled, and the form of the final report. Audits measure and communicate environmental goals, objectives, and improvement requirements to employees during the process. The audit also assures all employees that there is proactive management oversight and improves morale. The closing meeting brings together the preliminary findings and gives the site

personnel a chance to discuss these findings and possible solutions. The closing meeting allows the audited organization to provide input for incorporation in the final report.

7. *Public Relations and Marketing* — An audit program demonstrates a company's commitment to protect the environment and to reach sustainable development. Public recognition for such proactive measures is helpful in obtaining business certificates, such as licenses and permits, and corporations fare better with the community and the local government. Auditing assures an effective environmental program that is necessary for green marketing and the production of Green Reports. Consumers are environmentally conscious, e.g., public reaction to the Exxon Valdez oil spill in Alaska, and will act favorably to green companies. The use of auditing programs results in demonstrable business benefits.

8. *Securities and Exchange Commission* — The Securities and Exchange Commission (SEC) often inquires about information from environmental audits. Management of environmental liabilities is closely tied to corporate financial risk and company solvency. Since the stock values of corporations are sensitive to the annual financial report, public response to corporate performances is therefore closely tied to Environmental Management System and its performance monitoring. There is a move toward a requirement for environmental auditing results being reflected in the corporate financial statement. The Securities and Exchange Commission Required Disclosure table in Chapter 10 contains a list of the SEC's required disclosures.

9. *Proactive Management* — It has been demonstrated that it is cost-effective to know of noncompliance issues and to take the necessary actions proactively. If the noncompliance is discovered by outside agents, there is a possibility of being fined and the necessary corrective action must still be performed. The intervention by the agency may cause an unnecessary accelerated schedule and a greater expenditure than would have resulted from proactive management. The discovery by the agency may result in adverse publicity leading to third-party litigations, other agency involvement, and damage to the organization's public image. Such discovery leading to adverse publicity will result in a sharp drop in stockholder confidence. Examples of this effect include the following cases:
    - Chemical Waste Management — $900 million loss in stock value following closure of one of its landfills in California.
    - Union Carbide, Inc. — $700 million loss in stock value following the Bhopal incident.
    - Rollins, Inc. — $70 million loss in stock value following closure of its incinerator in Louisiana.

Having an effective, active auditing program in place can help to prevent such mishaps.

RISKS OF AUDITING

With all the above benefits there are some risks associated with an audit program. However, they can be managed properly to minimize risks to an operation. When planning an audit program or an audit, the following issues should be addressed and resolved as best as possible. You can then be aware of the type and degree of risk being undertaken.

1. *Cost* — The cost of an auditing program has been a major setback to many middle-sized companies, as well as to many large companies. Although the cost of the program can be offset by cost savings resulting from the program, the cost reduction is difficult to measure and usually is not. Many of the cost savings from a good audit program can be tangible in dollar figures. They include lower insurance premiums, fewer lost workdays, and increased production. Some of the intangible benefits are a better negotiating position with government agencies and unions and a more positive public image. More organizations should perform the cost-benefit analysis to better understand the justification of an audit program.
2. *Operations and Morale* — Temporary disruption of plant operations and the raising of employee concerns during an audit is unavoidable, but can be minimized by careful preplanning. Proper scheduling of interviews, use of nonintrusive observations, use of well-trained auditors, and the performance of an open, candid audit can minimize the disruption to plant operations. The audit process must be approached as an educational session for site personnel and presented as a site-personnel team effort. The audit team must emphasize the benefits of this collaborative team effort, allay employee concerns by being candid, and stress the overall benefits of the audit. Such a team effort will lead to site personnel taking ownership of their areas of responsibilities at the site. Often, the audit process and the findings are not explained to site personnel and thus create unnecessary concern.
3. *Disclosure* — The protection of audit findings is of critical concern. Disclosure can provide regulators and nongovernmental parties information that can lead to litigation. Excerpts from the report can be taken out of context and exaggerated in the media, leading to a negative public image. Therefore the control of the audit report, as well as the language in the report, are critical. The language of the audit report should be objective, factual, and nonjudgmental. The auditors need to be trained in writing reports in this manner. All audit reports should be reviewed by the legal department, and great

care should be exercised to protect the audit report under the attorney-client privilege. The preparation and distribution of the report should be made under the advice of legal counsel to ensure the sanctity of the attorney-client privilege.

If the court rules that the audit report is not protected by the attorney-client privilege (see Chapter 6), then the audit report can and may be used as an exhibit of the responsible corporate officers' willful and knowing violation of regulations. Depending on the litigants, there are two possible outcomes of disclosure. Any corrective action taken and demonstrable good faith effort can lead to favorable treatment in a civil or criminal litigation by enforcement agencies. However, a third-party litigant may sue for damages resulting from past actions, despite the good faith efforts to correct these past mistakes. Corporate officers do not want to be at the mercy of the court's discretionary decision and must carefully consider all the ramifications of disclosure. They should be fully aware of the disclosure and sentencing guidelines.

The regulated entities are, and have been, strongly advocating for fair treatment for voluntarily taking positive actions to assess and correct any past mistakes. In response to this pressing issue, the EPA has issued a new policy statement on self-evaluation and disclosure privileges. The statement of policy, *Incentives for Self-Policing: Discovery, Disclosure, Correction and Prevention*, was published in final form in the *Federal Register* on December 22, 1995 (Ref. 3). The EPA has stated in the policy that it will not seek gravity-based penalties for violations found through a voluntary environmental audit.

4. *Failure to Respond* — There is increased risk for liability when you fail to respond to audit results. Closure of findings from an audit is a critical component of any audit program. Failure to close non-compliance action items will lead to increased risk of liability. It is not expected that all non–United States Environmental Protection Agency compliances be corrected immediately; however, a reasonable, "good faith effort" is the key. Therefore a closure plan with adequate documentation that demonstrates the scheduled intentions and the good faith efforts, can alleviate the risks. When starting an audit program, the senior management should also be prepared to take corrective actions in reasonably good time.

5. *Prioritization* — It is difficult to conduct a consistent and systematic quantification of environmental issues for prioritization. The long-term issues are difficult to assess, and there are also ethical and esthetic issues that must be considered as well. Two facilities with similar environmental issues, in two different communities, will have different financial risks. How the two facilities prioritized their response to the noncompliance issue will depend on more than just

the technical and economic factors. Many organizations use scoring systems to prioritize their risk management issues. The parent organizations often have to respond to questions about the established scoring systems and have to be able to justify why a similar situation was ranked differently at two different facilities. The justification of the application of a scoring system is most often difficult to defend, but it is necessary for allocating resources to achieve overall maximum benefit for the company.

The prudent company will always double-check their environmental audit program, as implemented, to confirm that the program is meeting the company's standards on corrective action response, follow-up, confidentiality, and disclosure, ensuring proper risk management (Ref. 5).

4 FEDERAL AGENCY POSITIONS AND PERSPECTIVES

The federal agencies work together for enforcement and/or for sharing information. For example, the EPA and the DOJ have worked together on enforcement, and the EPA and the Securities and Exchange Commission (SEC) have shared information on disclosure. Federal and state agencies encourage voluntary audits and cooperate on enforcement actions; however, the states' legislation on self-evaluation privilege has caused the EPA and the DOJ to be concerned that the states' legislation will weaken federal enforcement programs. Fourteen states have passed this type of legislation (see Chapter 6).

At the same time, there is increasing regulatory demand for audits. The Federal Facility Compliance Act, October 6, 1992, has removed previous sovereign immunity on federal facilities and its employees. The EPA's Federal Facility Enforcement Office (FFEO) is responsible for enforcing the standards of the Act. The FFEO is also responsible for assisting facilities with compliance and has instituted a program that parallels the EPA's Common Sense Initiative program. This federal program uses strong enforcement combined with compliance assistance and promotes proactive technical programs, such as pollution prevention and environmental auditing.

Various federal and state legislations strongly encourage audits or make them mandatory.

- The United States Congress passed the Environmental Crimes Act of 1992, which requires courts to order a third-party audit of companies convicted of environmental crimes.
- The Senate's 1992 version of the proposed Clean Water Act Amendment reauthorization contains requirements for an audit as a prerequisite for discharge permit issuance for facilities subject to Superfund Amendments Reauthorization Act, Section 313.
- The proposed Senate Environmental Auditing Bill requires audits by a certified auditor for facilities regulated under the Clean Water Act, the Clean Air Act, the Safe Drinking Water Act, and the Solid Waste Disposal Act.

The environmental goal for our society is to encourage all the regulated parties to manage their environmental programs and liabilities in a socially conscientious manner. The goal of enforcement is to compel the integration of the environmentally progressive programs with business enterprises. A system that encourages, not forces, environmental progress is needed. The prosecution of violators is an enforcement tool that should only be used as a last resort. Given the identified risks and benefits of auditing, the fundamental questions then are, "Is the current federal policy fair?" and "Does the policy encourage regulated entities to perform self-evaluations?" Industry consensus on EPA's application of its 1986 Auditing Policy is that it has failed to provide this encouragement.

DEPARTMENT OF JUSTICE

The DOJ sentencing guideline, *Factors in Decisions on Criminal Prosecutions for Environmental Violations in the Context of Significant Voluntary Compliance of Disclosure Efforts by the Violator* (July 1991) is yet the strongest incentive for the use of voluntary audit programs. In the DOJ guidelines the following factors are normally considered without any guarantee of the disposition of the case:

- Voluntary, timely, and complete disclosure.
- Cooperation as far as the extent and quality of the violator's assistance to agency investigation.
- Scope of existing comprehensive environmental compliance program and corrective actions.
- Degree of violation through pervasive or systematic problem.
- Employee disciplinary action program for violating environmental compliance policies.
- Subsequent corrective actions with good faith.

Since this is only a guidance document, there is no guarantee that these criteria will be followed by the courts. Also note that the criteria apply to cases of criminal violations, not to civil cases. Furthermore, the protection of disclosure to the media is not mentioned in the DOJ guidelines. In the final analysis regulated entities have to choose between the risks associated with the use of an audit program and the risks associated with a violation discovery and the subsequent agency enforcement actions for noncompliance.

The House of Representatives (HR) has proposed a voluntary audit bill, HR 1047, the "Voluntary Environmental Self-Evaluation Act." The DOJ was asked to comment on the proposed legislation. The DOJ is in strong support of voluntary compliance, auditing, and disclosure. However, it is concerned that HR 1047 would "create an evidentiary and discovery privilege for 'voluntary environmental self-evaluations,' which appear to include any assessment, audit, investigation, or review carried out to determine compliance with environmental laws, no matter how informal." HR 1047 appears to grant immunity for any violation disclosed under an audit (Ref. 6).

The DOJ strongly opposes HR 1047, which would create evidentiary, discovery, and testimonial privileges for "voluntary environmental self-evaluation," and which would provide immunity from fines and penalties for violations that are the subject of a "voluntary disclosure" to a federal or state official, subject to specified exceptions. The DOJ considers evidentiary privilege, or immunity legislation, for self-evaluation and self-disclosure hopelessly flawed for the following reasons (Ref. 7):

- Immunity for self-disclosed violations would directly interfere with fair and effective enforcement of the law that protects the public's health and safety and our natural resources.
- An environmental audit privilege would impair law enforcement, result in abuse, conceal information vital to public health and safety, increase wasteful litigation, and interfere with the truth-seeking process.

The DOJ is of the opinion that there is no valid justification for an environmental audit privilege. There has not been any evidence of abusive use of audit information by criminal prosecutors in over 600 federal cases that were reviewed. In fact, the review concluded that environmental auditing is encouraged, not deterred, by strong enforcement, contrary to the claims of its proponents. A new evidentiary privilege would only increase the involvement of lawyers, and it is unlikely to generate more environmental audits or improve environmental compliance (Ref. 7).

To comprehend the legal conflict, it is essential to understand the distinction between *privilege*, which is protection by not being required to release information, and *immunity*, which is protection from penalty after revealing one's transgressions. Looking at enforcement strictly from an institutional philosophy and practice, the privilege and immunity being requested by the regulated community seem outrageous. Voluntary disclosure of flagrant violations and immunity from penalties would only encourage more misconduct. By granting immunity for environmental misconduct, environmental enforcement activities will be hampered. This punishes the law-abiding regulated entity and rewards those who willingly violate the law. The fairness of the punishment to those who have broken the law and caused damage is still under question by the DOJ.

Two critical issues to law enforcement and to the ability of the public to protect itself from environmental threats are being challenged. There are good and bad environmental actors out there, and the justice system cannot be designed just for the good actors; usually it is the reverse. By allowing violators to shield environmental problems from the public, granting immunity for environmental misconduct, and tying up environmental enforcement in endless litigation, this bill would punish the law-abiding, environmentally sound company and reward those who flout the law (Ref. 7).

The DOJ holds a similar position on the recent state legislation that protects audit information with immunity privileges. It is similarly cautious about the risk of weakening enforcement programs and of the immunity causing long and extended litigation. Many state and local officials concur with the DOJ in opposition to audit privilege/immunity bills. However, 14 states' legislative bodies in the past 3 years have either considered or passed legislation for audit privilege/immunity.

ENVIRONMENTAL PROTECTION AGENCY

In 1994 the Office of Enforcement and Compliance Assurance (OECA), in response to Administrator Carol M. Browner, began to determine whether additional incentives were needed to encourage voluntary disclosure and correction of violations uncovered during environmental audits. An environmental auditing policy statement was previously issued in 1986. Currently, there are no regulations requiring an audit program, but as a result of the OECA's determination, a new EPA policy statement (1995) was issued. The new policy enhances the 1986 statement, strongly encourages voluntary self-evaluation programs with specific inclusion elements, and provides a more level playing field, however, without providing protection from disclosure.

Environmental Auditing Policy Statement of 1986

The 1986 policy statement was issued in three parts: a preamble, a general EPA policy on environmental auditing, and an appendix. A brief overview of the 1986 policy is below. A copy of the full text of this policy may be found in Appendix A.

In its policy the EPA encouraged the use of environmental auditing by regulated entities to help achieve and maintain compliance with environmental laws and regulations, as well as to help identify and correct unregulated environmental hazards.

The EPA also encouraged federal agencies, subject to environmental laws and regulation, to develop and maintain an auditing program to supplement EPA and state inspections. The program that the federal agencies developed and any subsequent audits would be treated in the same manner as other regulated entities.

The EPA also recognized the independent authority of the state or local governments having jurisdiction over regulated entities within their boundaries. The EPA encouraged the governments to adopt EPA's or a similar policy to advance the use of auditing in a consistent manner. The agency considered an effective state/federal partnership was needed to accomplish the goal of achieving and maintaining high levels of regulatory compliance. While nothing in the policy preempted the states and local governments from developing

FEDERAL AGENCY POSITIONS AND PERSPECTIVES

their own approaches to auditing, the EPA did want them to consider five basic principles during their process. In brief, the basic principles are as follows:

- Regulated entities should continue to report or record compliance information required under existing law. Information cannot be withheld just because it was generated by an audit.
- Regulatory agencies cannot make promises to limit or forego enforcement actions in exchange for use of an auditing system.
- Inspection priorities for regulatory agencies should focus on compliance performance and environmental results.
- Regulatory agencies must continue to meet minimum program requirements.
- Regulatory agencies should not attempt to prescribe the precise form of the other regulated entities' environmental management or auditing programs.

EPA Recommendations on Auditing Programs — The EPA supports and encourages the use of auditing as a tool for quality assurance and quality control to achieve and maintain environmental compliance, as well as to correct unregulated hazards. "Environmental audits evaluate, and are not a substitute for, direct compliance activities such as obtaining permits, installing controls, monitoring compliance, reporting violations, and keeping records," according to the 1986 policy. The EPA does not intend to mandate auditing, except for those requirements that may be part of a settlement agreement. Audits should remain voluntary since this practice has been widely accepted and the audit quality depends on staunch management support of the auditing program and goals.

EPA considers that an effective environmental auditing program should include the following elements:

- Explicit top management support for auditing and commitment to follow-up on audit findings.
- An auditing function independent of operational activities, usually as a corporate function or outside consultant.
- Adequate team staffing and auditor training.
- Explicit audit program objectives, scope, resources, and frequency.
- A process that collects, analyzes, interprets, and documents information sufficient to achieve audit objectives. The information collected should be sufficient, reliable, relevant, and useful to support the audit findings and recommendations.
- A process that includes specific procedures to promptly prepare candid, clear, and appropriate written reports on audit findings, corrective actions, and schedules for implementation.
- A process that includes quality assurance procedures to ensure the accuracy and thoroughness of the audit.

Specific Auditing Issues — The EPA will not *routinely* make requests for audit reports; however, it does reserve this right when the report is needed for criminal investigation. Although the EPA will not promise to forgo inspections, reduce enforcement actions, or offer other such incentives in exchange for implementation of auditing and other sound environmental management practices, it will take into account the efforts of a regulated entity to audit and to rectify incidences. The EPA's authority to "request an audit report or portions thereof, will be exercised on a case-by-case basis." The EPA will not request the whole audit document when only the relevant portion will be sufficient for its purposes. The EPA will make these requests in cases where they determine the following:

- The report is needed to accomplish statutory mission, and information is material to a criminal investigation.
- The audits are part of a consent decree or settlement agreement.
- Part of the report is already reportable or otherwise accessible to the EPA.

This is a list of *examples only* and is *not* all inclusive.

The EPA will not forgo inspections, reduce enforcement, or offer other incentives in exchange for the implementation of an auditing program, as the agency is required by law to independently assess the status of compliance at facilities. However, it will take into account a facility's audit provisions when considering enforcement settlements.

The EPA may impose audit provisions in consent decrees and other settlements where auditing could provide a solution for and reduce the likelihood of reoccurrence of identified problems. These provisions are most likely to be used when a pattern of violation can be attributed to poor functioning or an absent environmental management system, or if there is an indication that similar noncompliance problems may exist at other facilities. The types of audits that the EPA considers for settlements are compliance audits, management audits, or both. The compliance audit is an independent assessment of a facility's compliance with the applicable laws and regulations, and the management audit is an independent assessment of a facility's environmental compliance policies, practices, and controls.

Interim Policy Statement on Self-Policing

The new policy was developed in close consultation with the EPA's regional offices and the DOJ and with input from the general public, the private sector, and government agencies. An empirical, information-gathering, open public meeting was held July 27–28, 1994, in Washington, D.C. The meeting was well attended, and the proposal to change the EPA policy received a great deal of response from stakeholders from industry, trade groups, state environmental commissioners, attorneys general, public interest groups, and professional auditors. A second public meeting was held in San Francisco, January 19–20, 1995.

In announcing the advent of the public meetings the EPA took the opportunity to issue a reassessment of its policy on auditing. The notice in the *Federal Register* July 28, 1994, was entitled *Clarification on Policies Related to Environmental Auditing* (59 FR page 38455). A copy of the full text of this policy reassessment may be found in Appendix A.

The interim policy intended to promote improved environmental compliance by clarifying the EPA's position to voluntary self-evaluation, voluntary disclosure, and prompt correction of violations. The interim policy provided incentives that eliminated or substantially reduced the gravity of civil penalties and stated that the agency would not refer cases for criminal prosecution when specified conditions were met.

The EPA also recognized that states are important partners in federal enforcement and that the EPA is required to establish a certain minimum consistency in federal enforcement so that the sanctions a business faces for violating federal law do not depend on where the business is located.

The EPA believes that the best way to achieve compliance with federal laws that protect public health and safeguard the environment is through voluntary cooperation from businesses and other regulated entities. Instead of a punishment-oriented program for those caught while trying to fix past problems, the emphasis is on an incentives program for those taking responsibility for voluntary auditing, disclosure, and correction of violation. This provides a level playing field so that violators do not obtain an unfair economic advantage over their competitors who made the necessary investment in compliance. Penalties also promote protection of the environment and public health by adoption of pollution prevention and recycling programs.

Other incentives for voluntary auditing appeared in an EPA notice in the *Federal Register* (FR) on June 21, 1994 (59 FR, page 32062), requesting proposals for pilot projects under the new Environmental Leadership Program (ELP). The ELP pilot projects were to test criteria for auditing and certification of voluntary compliance programs. A successful ELP program would allow companies to gain public recognition and reduce inspections by the agency. This was a step in the right direction.

The EPA also recognizes global voluntary incentives for the use of audit programs, such as ISO 14000. ISO 14000 will provide a powerful model to improve audit programs' overall compliance and to move from compliance audits to environmental management system (EMS) audits. The EPA has submitted comments on the ISO 14000 standards and will continue a dialogue with this and other parties. The American National Standards Institute is the United States representative to the ISO 14000 committee and eventually will adopt these standards.

Environmental Leadership Program — The EPA had been developing the ELP parallel to the development of the new policy. The EPA used the observations of these programs to help form the interim and final policy statements on self-policing.

The genesis of the EPA ELP in April of 1992 was to complement the Occupational Safety and Health Administration's Voluntary Protection Program (VPP) in recognizing those environmental programs that go beyond mere regulatory compliance. The ELP was designed to be a voluntary excellence program. It is one of President Clinton's strategies in reinventing the environmental regulations that look at superior programs that exceed existing regulatory compliance. The ELP looked at auditing programs, pollution prevention integration, EMS, public accountability, and the mentoring of small businesses. The ELP complements the Common Sense Initiative and Project XL. The Common Sense Initiative focuses on specific industry sectors, while the ELP and Project XL are facility-specific projects. The ELP is designed to examine innovative approaches to compliance within the existing regulations. Project XL is different in that it includes flexibility from existing regulation in exchange for the attainment of environmental results beyond achievement through compliance of those regulations.

The purposes of the pilot phase of the ELP are to examine the basic components of what should be state-of-the-art compliance management systems, to identify the verification procedures that will ensure that the ELP is working, to establish measures of accountability so that the systems will be credible to the public, and to promote community involvement in understanding and supporting innovative approaches to compliance. The ELP pilot projects were set up to test one or more of the six basic ELP criteria that were identified by EPA as parts of the basic components. The basic criteria are as follows:

- Environmental management systems — The design of management systems that both prevent violations and ensure continuous environmental improvement.
- Multimedia compliance assurance — Developing and sharing multimedia inspection protocols that can be integrated into the corporate EMS and into daily facility management practices.
- Third-party verification and self-certification — Using independent verification of compliance audit methods to assure accuracy of self-audit results and methods, and developing protocols for the establishment of self-certification reporting requirements in company operations where there is a relatively low environmental risk, e.g., daily operations, routine periodic reporting.
- Public accountability — Enhancing the organization's accountability to the public through the creation of performance measures that are accurate and meaningful to the public.
- Community involvement — Initiating involvement through cooperation between facilities and local communities to negotiate environmental goals and track progress in meeting them.

FEDERAL AGENCY POSITIONS AND PERSPECTIVES

- Mentoring — Developing the concept of guidance in which large or more sophisticated companies develop model training programs to help small businesses stay in compliance while remaining competitive.

On April 7, 1995, the EPA announced in the *Federal Register* the 12 ELP pilot projects that had been chosen for study to test the design of the ELP program components. The 12 projects made up of 15 industry sites and federal installations were selected from 40 volunteers. The pilot projects began on May 1, 1995, and are scheduled to be completed on April 30, 1996. The pilot project participants benefit from public recognition, limited grace period[4] to correct any violation, and participation in a joint initiative to identify methods for reducing agency inspections and reducing paperwork.

The EPA, the state, and the corporate host will use the 12-month period of the pilot phase to test key concepts and evaluate results. At the end of the pilot phase EPA will refine the program and make its benefits broadly available in early 1997 to those willing to meet its criteria. The end result will be an EPA-recognized, compliance management system with certification opportunities for industries and federal facilities. Regulated entities with a certified compliance management system will be able to reduce unnecessary inspections at their facilities. ELP is important to the auditing procedure because it will shape how auditing should be performed and documented.

The pilot projects have taken the existing practices set forth by ISO 14000, National Sanitation Foundation 110, British Standard 7550, European Community Management and Auditing Scheme, Global Environmental Management Initiative, and Coalition for Environmentally Responsible Economics to develop a state-of-the-art certifiable EMS program.

The negotiated scope for each project is different and the 12 projects represent a broad spectrum of our nation's industry: manufacturing, chemicals, waste management, computers, energy, and printing. The intent of the ELP was to look at the state of the art compliance management systems, to look at the basic components, to assure a successful check and balance system, to develop and test a verification process that will assure the public and the enforcement agencies that programs are working, to develop and test the concepts of third-party auditing and self-certification for compliance, and to develop measures of accountability so that the compliance management systems will be credible to both the public and the enforcement agencies. Community and employee involvement is a key component of the pilot projects, and the 12 projects are summarized in Table 4.1.

Compliance Audit Guidance — The ELP Compliance Audit Guidance (currently in draft stage) looks at both auditor qualifications and audit activities

---

[4] The period is for 90 days, but does not apply to violations that are criminal, that may present an imminent and substantial endangerment to public health or the environment, for which a prior enforcement response has been taken, or where an agency determines that significant economic benefit has been realized.

Table 4.1 Summary of the Environmental Leadership Program Pilot Projects

Organization and location(s)	Product/function	Pilot project description
Arizona Public Service Deer Valley Facility Phoenix, AZ	Electric power generation	Proposed use of an EMS audit
Ciba-Geigy St. Gabriel, LA	Crop protection and speciality chemicals, and textile dyes	Three projects: EMS; multimedia compliance assurance; and community involvement
Duke Power Riverbend Steam Station Mount Holly, NC	Electric power generation	Develop and test one or more of the six primary ELP criteria
The Gillette Company Boston, MA (South) Chicago, IL (North) Santa Monica, CA	Consumer products	Develop a monitoring system using independent third-party auditors for verification of having implemented a necessary EMS
The John Roberts Co. Minneapolis, MN	Reports, catalogs, direct mail pieces	Mentoring of smaller businesses and sharing auditing tool information
McClellan Air Force Base Sacramento, CA	High-tech industrial repair facility	Developing and sharing multimedia inspection protocols for integration into a corporate EMS to enhance public accountability and community involvement
Motorola Inc. Oak Hill Facility Austin, TX	Semiconductor and wafers	Mentoring and a total quality EMS program

Table 4.1 Summary of the Environmental Leadership Program Pilot Projects (Continued)

Organization and location(s)	Product/function	Pilot project description
Ocean State Power Burriville, RI	Independent electric power generation	Develop a mentoring program to provide environmental, safety, and health management assistance to smaller companies, and also demonstrate its EMS
Puget Sound Naval Shipyard Bremerton, WA	Naval shipyard and industrial facility	Use pollution prevention opportunity assessments in beyond compliance audits in partnership with state and federal agencies
Salt River Project Tempe, AZ	Public power utility and water supplier	Three projects: Clean Air Act Title V EMS; self-certification; and customer environmental awareness training
Simpson Tacoma Kraft Co. Tacoma, WA	Packaging materials	Tiered approach to beyond compliance auditing and a technical exchange program
WMX Technologies, Inc. Arlington, OR	Waste management, disposal, and recycling	Demonstrate cooperation with regulatory agencies to achieve a high standard of compliance, and shared information about an innovative EMS

during the previsit, on-site, and postvisit phases of the audit. When selecting an auditor, the following factors should be considered:

- Independence from the facility to perform an objective audit.
- Knowledge, skills, education, and experience commensurate with the size and complexity of the specific facility being audited.
- Level of professional responsibility to conduct a highly effective audit.

Previsit activities should conclude with the auditors understanding the facility-wide environmental issues and the basic industrial processes and operation and being able to identify all substances or waste material that could be released into the environment, as well as being familiar with community concerns and sensitive environments in the proximity to the facility.

On-site activities should recommend sampling and analysis for situations where data are lacking. The recommendation would be appropriate when waste analysis data are incomplete, unknown waste is found, or there are unexplained, suspicious stains or discolorations in waste management areas, etc. Where a situation calls for on-site sampling and the auditors do not implement verification sampling, rationale for this decision should also be provided in the audit report. Beyond compliance activities, such as pollution prevention, waste minimization, unregulated risks, and the company's environmental policy, should be included in the audit protocol.

The postaudit conference should end with a mutual (facility and auditor) understanding of what is to be expected in the audit report and the expected date of distribution of the report. The postvisit activities includes delivery of the final report, which includes the audit protocol used, in a reasonable period. The observations and interview information acquired should be discussed along with the on-site sampling data and previously existing data.

The facility is responsible for developing a corrective action plan that includes all areas of noncompliance. The plan must include the activities necessary to correct the problem, a schedule of implementation and compliance attainment, any activities already implemented to address risks identified, and where applicable, a discussion of source elimination/reduction efforts to address noncompliance.

Self-Policing Policy of 1995[5]

On December 22, 1995, the EPA announced the *Incentives for Self-Policing: Discovery, Disclosure, Correction and Prevention of Violations*, in the *Federal Register* (Ref. 4). Under the new policy, the agency will greatly reduce civil penalties and limit liability for criminal prosecution for regulated

---

[5] This section of Chapter 4 was provided by Mr. Brian Riedel (see profile preceding Chapter 1).

entities that meet the policy's conditions for discovery, disclosure, and correction.[6]

Policy Incentives — Under the policy, the EPA will not seek gravity-based[7] civil penalties for violations that are discovered through an EMS or an environmental audit and that are promptly disclosed and expeditiously corrected, provided the other policy conditions are met. Where violations are discovered by means other than an EMS or an environmental audit, but are promptly disclosed and expeditiously corrected, the EPA will reduce gravity-based penalties by 75% provided the other policy conditions are met. The agency will generally not recommend to the DOJ that criminal charges be brought against entities that meet all of the policy conditions. Finally, the policy restates EPA's policy and practice of not routinely requesting environmental audit reports.

Safeguards — While the 1995 policy contains significant incentives for encouraging discovery, disclosure, and correction of violations, it also contains very important safeguards to deter irresponsible behavior and protect the public and the environment. For example, the policy requires entities to take steps to prevent recurrence of the violation and to remediate any harm caused by the violation. In addition, the policy does not apply to violations that resulted in serious actual harm or may have presented an imminent and substantial endangerment to human health or the environment. Moreover, entities are not eligible for relief under the policy for repeated violations. The policy does not apply to individual criminal acts or corporate criminal acts arising from conscious disregard or willful blindness to violations. Finally, the EPA retains its discretion to collect any economic benefit gained from noncompliance to preserve a "level playing field" for entities that invest in timely compliance.

Incentives and Behavior — The self-policing policy provides additional incentives for entities to utilize the critical compliance tools of environmental auditing and compliance management systems. These incentives add to the many existing business reasons for entities to develop and maintain environmental auditing and compliance management systems.

In 1986 the EPA announced that it was the agency's policy to encourage environmental auditing as a means to help achieve and maintain regulatory compliance. In the 1995 policy EPA's tactics had evolved toward encouraging the use of compliance tools, such as auditing and management systems, by providing penalty incentives and a safe harbor[8] from criminal prosecution. It is important to recognize that this evolution is likely to continue as organiza-

---

[6] A copy of the policy and its comprehensive preamble appears in Appendix A.

[7] The "gravity" component of a penalty represents the "seriousness" or "punitive" portion of penalties. The other major part of a penalty, the economic benefit component, represents the economic advantage a violator gains through its noncompliance.

[8] Safe harbor, in a legal sense, is the establishment through regulation of legal protective measures and privileges.

tions develop more effective tools to manage the environmental aspects and impacts of their activities, services, and products.

Policy Conditions — There are nine conditions that a regulated entity must satisfy for the EPA not to seek gravity-based penalties. The conditions are as follows:

- Systematic discovery
- Voluntary discovery
- Prompt disclosure
- Discovery and disclosure
- Correction and remediation
- Prevent recurrence
- No repeat violations
- Other violations excluded
- Cooperation

Systematic Discovery — Entity must discover the violation through an environmental audit or environmental management system to obtain full gravity penalty mitigation and criminal safe harbor. The 1995 policy provides full mitigation of gravity-based civil penalties and a criminal safe harbor for entities that discover violations through an environmental audit or EMS reflecting due diligence, provided the other policy conditions are met.

The policy provides relief to entities that discover violations through an objective, documented, systematic procedure or practice reflecting the regulated entity's due diligence,[9] in preventing, detecting, and correcting violations, provided the other conditions are met. The due diligence criteria in the 1995 policy includes the following:

- The development of compliance policies, standards, and procedures to meet regulatory requirements.
- Allocation of responsibility to oversee conformance with these policies, standards, and procedures.
- Mechanisms, including monitoring and auditing of compliance, and the EMS to ensure the policies, standards, and procedures are being carried out.
- Training to communicate the standards and procedures.
- Employee incentives to perform in accordance with the compliance policies, standards, and procedures.
- Procedures for the prompt and appropriate correction of violations, including program modifications needed to prevent future violations.

---

[9] Due diligence is defined as systematic efforts meeting criteria based on the 1991 United States Sentencing Commission Sentencing Guidelines Manual, Chapter 8, Sentencing of Organizations, Part A — General Application Principles (effective November 1, 1991).

The inclusion of EMSs in the policy represents a very positive and significant revision to the interim auditing policy. Stakeholder written and oral comments indicated that ongoing, comprehensive, and systematic efforts to prevent, detect, and correct violations should be rewarded at least as much as environmental auditing. The difference between a compliance audit and an EMS is similar to the difference between a snapshot and a video.

It is also very significant that the EPA may require, as a condition for penalty mitigation, that a description of an entity's EMS be made public. This type of public disclosure has the potential to push the state-of-art EMS and encourage the use of the EMS as a benchmark among suppliers and competitors. The public availability of the EMS can also provide valuable information for insurers, financial markets, investors, and lenders, providing a basis for market-based incentives.

Voluntary Discovery — The policy applies to all violations except those discovered through mandated monitoring or sampling requirements. A significant revision made to the interim auditing policy is the elimination of the distinction between violations that are required to be reported and violations that are not required to be reported. To provide maximum opportunity to encourage compliance, and to do so without sacrificing the integrity of critical reporting systems, the policy provides relief on voluntary disclosure of all violations except those violations discovered through mandated monitoring or sampling requirements, provided the other policy conditions are met. Examples of violations not covered by the policy include emissions violations detected through continuous emissions monitoring, violations of National Pollution Discharge Elimination System permits detected through required monitoring, or violations discovered through a compliance audit required to be performed by the terms of a settlement agreement.

Prompt Disclosure — The regulated entity promptly discloses the violation in writing to the EPA. Under the policy the entity must fully disclose in writing to the EPA that a violation has occurred or may have occurred, within ten days after discovery. The inclusion of the "may have occurred" language recognizes that in situations where the entity is unsure whether a violation had occurred, it is best for the entity to disclose the potential violation to the EPA for a definitive determination. The EPA may accept disclosures more than ten days after discovery if more time is needed to make a compliance determination of a complex violation and circumstances do not present a serious threat.

Discovery and Disclosure — The entity must disclose the violation prior to imminent discovery by the government. The entity must identify and disclose the violation before the government has discovered or will discover the violation. Thus the entity must disclose the violation prior to commencement of a government inspection or investigation, issuance of an information request, notice of citizen suit, filing of a third-party compliant, or reporting by a whistle-blower.

Correction and Remediation — The entity must expeditiously correct the violation and remedy harm. The entity must correct the violation expeditiously and within 60 days, certify correction, and take appropriate measures to remedy any harm caused by the violation. If more than 60 days is needed to correct the violation, the entity must notify the EPA before the 60-day period has passed. Where appropriate, the EPA may require a written agreement to satisfy requirements for correction, remediation, or prevention measures, especially where such measures, are complex or lengthy.

Prevent Recurrence — The regulated entity must agree to take steps to prevent recurrence of the violation. The entity's efforts to prevent recurrence of the violation may involve modifying its environmental auditing program or EMS.

No Repeat Violations — The violation had not occurred at the same facility within the past three years and was not part of a pattern of violations at the parent company within the past five years. The policy does not apply to repeat violators. The EPA has established a method to determine when repeat violators should not be eligible for relief. Under the policy, relief is granted only if the same or closely related violation has not occurred at the same facility within the past three years, or is not part of a pattern of violations at the facility's parent organization within the past five years. This policy exclusion provides entities with a continuing incentive to prevent repeat violations and avoids the unfairness of granting relief repeatedly for the same or similar violation.

Other Violations Excluded — The violation is not one that resulted in serious actual harm or may have presented an imminent and substantial endangerment, or does not violate the specific terms of an order or agreement. The policy does not apply to violations that resulted in serious actual harm or may have presented an imminent and substantial endangerment to human health or the environment. Coverage of the policy to such violations would undermine deterrence and reward regulated entities for delinquent management of its environmental activities. The policy also does not apply to violations of the specific terms of any settlement agreement.

Cooperation — The entity must cooperate with the EPA. At a minimum, the entity must provide the necessary information requested by the EPA to investigate the violation, and any noncompliance problems and environmental consequences related to the violation.

Relationship to State Laws, Regulations, and Policies — The EPA remains opposed to environmental audit privileges that provide a cloak of secrecy over evidence of environmental violations and that are in conflict with the public's right to know. The EPA also remains opposed to blanket immunities or amnesty for violations that reflect criminal conduct, present serious threats or actual harm to the environment, allow noncomplying entities to gain an economic advantage, or reflect a repeated failure to comply with federal law. The EPA will work with states to encourage their adoption of policies that reflect the incentives and conditions outlined in the 1995 policy and restates its pledge to work with states to address any provisions of state audit privilege or penalty immunity laws that are inconsistent with the policy.

Applicability — The policy supersedes any inconsistent provisions of the EPA's media and program-specific penalty and enforcement response policies. It will operate in conjunction with other EPA enforcement policies to the extent they are consistent with each other. However, a regulated entity may not receive additional penalty mitigation for having met the same or similar conditions under other enforcement policies. This policy will also not apply to violations that have received penalty mitigation under other enforcement policies. The policy is intended to be utilized for settlement of administrative and judicial enforcement actions, not for pleading purposes.

Tracking of Cases — The EPA plans to carefully track cases handled under the new policy and make information about these cases publicly available. This will provide interpretative guidance, help ensure that the policy will be applied consistently, and instill confidence in the policy.

Industry Concerns — The policy addresses three general problem areas raised by the regulated community. First, the EPA's penalty response to self-discovered and self-disclosed violations (under its approximately 24 media and program-specific enforcement policies) did not provide the consistency and certainty of enforcement response that regulated industry sought. The final policy establishes a multimedia, consistent, and certain enforcement response for those entities that discover, disclose, and correct environmental violations and meet the other safeguards under the policy.

Second, the regulated community perceived that the EPA's enforcement policies placed entities that proactively seek to identify and disclose noncompliance in no better position, or even in a worse position (exposure to criminal and civil liability), than regulated entities that have not made the efforts to identify noncompliance. The discovery of violations triggered additional reporting requirements and potential criminal liability "for knowingly or intentionally failing to report." The audit report could be used as a "road map" to investigate and prosecute violations revealed in the report.

The EPA believed that perceptions surrounding these concerns were not supported by the facts. For example, the EPA and DOJ were not able to identify a single federal or state criminal prosecution of a regulated entity for violations uncovered through an audit and self-disclosed before an independent government investigation was under way. Moreover, a 1995 Price Waterhouse survey indicates that the EPA policies and practices have not discouraged auditing. The respondents that do not audit generally do not perceive any need to audit, and concern about confidentiality of audit information is one of least important factors in their decisions not to audit. In fact, corporate respondents indicated that the EPA activity contributed to their decision to audit.

Nonetheless, through this policy the regulated community will have a criminal safe harbor and civil penalty mitigation to further encourage environmental auditing and the development and implementation of systems to prevent, detect, and correct environmental violations. However, regulated entities that are not prepared to promptly disclose and expeditiously correct discovered violations are not protected by the policy and may be running a substantial risk. Overall, the policy removes perceived nonincentives to discover, disclose, and correct environmental violations. In addition, the 1995 policy restates the EPA's practice and policy since 1986 of not requesting or using an environmental audit report to initiate a civil or criminal investigation of the entity.

Finally, industry was concerned that criminal acts of "rogue employees" could inculpate the corporation and corporate officers, where the corporation and individual corporate officers are not otherwise culpable. The policy makes it clear that as long as all of the nine conditions are satisfied, the EPA will not recommend that criminal charges be brought against the regulated entity. This criminal safe harbor is available so long as the violation did not involve a prevalent philosophy or practice to condone or conceal the violations, or a high-level corporate or managerial conscious involvement in or willful blindness to the violations. The EPA reserves the right to recommend prosecution for the criminal acts of individuals. The EPA has not referred a criminal case for prosecution of corporate officers, nor has the DOJ criminally prosecuted corporate officers solely on the basis of the corporate officer's position in the company.

Privileges and Immunities — In an effort to address some of the perceived concerns regarding government and third-party use of audit information, some in the regulated community have turned to state and federal legislation. Since June 1993 14 states have enacted legislation to create evidentiary privileges for environmental audits. Some detractors have referred to these laws as "environmental secrecy acts."

Nine states (including eight of the "privilege" states) have passed penalty immunity or amnesty provisions for self-disclosed violations discovered through an audit. These penalty immunity provisions vary widely in terms of the extent of exceptions for criminal behavior, serious harm or threats of harm, recovery of economic benefit, and repeat violations. Some detractors have referred to these provisions as "confess and forgive" laws. Federal audit privilege and penalty immunity bills have been introduced in both houses of Congress. However, the EPA still maintains its opposition to this type of immunity and audit privilege legislation.

AUDITING AND ENFORCEMENT AT FEDERAL FACILITIES[10]

Status of Environmental Auditing and Enforcement at Federal Facilities

The United States government is clearly one of the most important sectors of the economy for which the EPA has environmental regulatory responsibility. The federal government owns nearly one-third of all the land in the United

---

[10] This section of Chapter 4 was provided by Mr. Richard Satterfield (see profile preceding Chapter 1).

States and spends nearly $12 billion per year on environmental issues. By comparison, the annual EPA budget is only about $7 billion. The Department of Defense alone is engaged in cleanup at 1800 military installations in the United States and overseas. Over 90 of these installations are on the Superfund[11] National Priorities List.

Historically, the EPA has been cautious in advising regulated entities on the use of environmental audits and has generally pursued an informal policy of "encouragement and assistance" on the matter. The EPA publicly addressed environmental auditing for the first time in 1986 when the agency released its first environmental audit policy in the *Federal Register* on July 9, 1986 (Ref. 3). The policy encouraged all regulated entities to use environmental auditing to help achieve and maintain compliance with environmental laws and regulations. The 1986 policy also contained explicit language to encourage federal agencies to develop environmental audit programs, stating that "to the extent feasible, the EPA would provide technical assistance to help federal agencies design and initiate audit programs."

In 1987 the EPA's Office of Federal Activities conducted a survey of audit activities at federal facilities. The results of the survey showed that larger federal agencies, such as those within the Department of Defense and the Department of Energy, had begun to develop audit programs; however, for other federal agencies, environmental auditing and audit programs were either nonexistent or were "work in progress." In 1988 the EPA sponsored a nationwide environmental auditing conference for federal agencies and in 1989 issued guidelines, including a generic environmental audit protocol, to assist federal agencies in establishing audit programs.

A comprehensive report by the Government Accounting Office (GAO) entitled, *Environmental Auditing; A Useful Tool That Can Improve Environmental Performance and Reduce Costs*,[12] released in April 1995, represented the conclusion of an 18-month study that reported on the experiences of both private and federal agencies in reducing liabilities by performing environmental audits. The report also revealed the extent to which federal agencies use environmental auditing and the benefits that could accrue from its wider use.

The report also described a GAO study of "benchmark organizations" that revealed that environmental auditing has a substantial positive effect on lessening liability resulting from regulatory enforcement, environmental impairment, and third-party exposure/complaint. One of the principle findings of the study stated that environmental auditing activity is rare at most federal agencies and is particularly diminished at smaller federal agencies that are typically understaffed with sufficient technical expertise and therefore do not perform

---

[11] Superfund is the Comprehensive Environmental Response Compensation and Liability Act (CERCLA).

[12] *Environmental Auditing: A Useful Tool That Can Improve Environmental Performance and Reduce Costs*, Government Accounting Office, Publication Number, GAO/RCED-95-37.

as well as the Departments of Defense and Energy in terms of environmental management. The EPA refers to these agencies as Civilian Federal Agencies.

The report findings specifically state that Civilian Federal Agencies' senior management find little encouragement to implement audit programs because of the nonincentives contained in the EPA's 1986 auditing policy and the fact that the EPA rarely inspects their facilities. The GAO also concluded that environmental auditing expertise at Civilian Federal Agencies was lacking and that, because the management has shown little interest in implementing audit programs, funds for developing expertise via training were insufficient. The report also indicated that federal agencies are further discouraged from performing audits because of several incidents whereby the EPA regions have requested audit reports for reasons other than those allowed under the current (1986) EPA audit policy. The GAO also cited the EPA's 1986 audit policy with being too limiting in assuring that it will reward self-discovery and reporting of violations.

The GAO has determined that the EPA needs to create more incentives for increasing environmental audit activity at Civilian Federal Agencies by the following means:

- Enforcing the EPA's current stated policy of limiting requests of audit reports by personnel in the EPA regions
- Changing the EPA's existing audit and enforcement policies to encourage regulated entities to perform more environmental audits
- Providing sustained technical assistance to the agencies
- Providing a greater show of enforcement at Civilian Federal Agencies facilities throughout the EPA regions

The EPA concurred with the majority of the findings and recommendations contained in the GAO report. However, the EPA also provided written response to those items that called on the EPA for further action. In its response the EPA stated that while it acknowledged the agency could and should do more to promote the development of federal agency audit programs, federal agency managers and executives must also demonstrate top management support for developing audit programs by ensuring that sufficient expertise for the management of an audit program is developed from within their own agency.

The federal government must, by law, meet the same environmental standards as those in the private sector, but federal facility compliance rates have traditionally been lower than the private sector. President Clinton made clear his views on the subject in his 1993 Earth Day message: "[It] is time that the United States government begins to live under the laws it makes for other people." In addition to the President's views, the Congress and the American people have also provided strong messages of their expectations for the various agencies and departments of the government in regard to environmental performance. Over the last decade the role of federal facilities in pollution control and abatement has been scrutinized by Congress and the media. As such, Congress has modified existing laws and enacted new ones to require federal agency compliance with environmental requirements. Perhaps the most significant of these is the Federal Facilities Compliance Act (FFCA), which requires that federal facilities comply with all applicable federal, state, and local requirements concerning the treatment and disposal of solid and hazardous waste. This requires federal agencies to comply with their procedural provisions, including penalties for violations. Prior to passage of the FFCA, the EPA took Resource Conservation and Recovery Act (RCRA) enforcement actions against federal agencies differently than against private parties, negotiating Federal Facility Compliance Agreements to bring federal facilities back into compliance. In passing the FFCA in 1992, Congress clarified that administrative order authority has been given directly to the EPA administrator. The waiver of sovereign immunity found in RCRA was amended by the 1992 act. Congress further clarified in the act that federal agencies are "persons" for the purpose of RCRA. Therefore the EPA now has administrative compliance order authority against federal facilities. The act also includes a strong waiver of sovereign immunity and thus allows the imposition of stiff fines and penalties for federal facilities found to be out of compliance by the EPA or state inspectors. Recent modifications in federal and many state criminal enforcement laws and policies are likely to encourage federal prosecutors to consider whether violations are criminal offenses and to target officials responsible for environmental crimes.

Increasingly, federal agencies have been pressured to develop sound environmental management programs that will ensure their compliance with environmental requirements. The task of maintaining environmental compliance is daunting with over 11,000 pages of codified regulations spanning a full range of multimedia issues. Moreover, a recent signing of several presidential executive orders has placed expectations on federal agencies to include pollution prevention goals as part of their overall plans for environmental management. An environmental management program that is purely reactive to these manifold requirements is a serious handicap for federal agency personnel and their managers.

In all this, the role of the federal agency environmental compliance officer is crucial. The officer must monitor the agency's compliance with environmental statutes and regulations and create and oversee effective programs that promote compliance, protect human health, and preserve the environment. Basic to the performance of the compliance officer's responsibilities is the enabling systems or infrastructure within an agency that support environmental compliance as the foundation of an effective environmental compliance program. These enabling systems should cut across all planning, organizing, implementing, and measuring processes within the agency.

To prevent environmental problems from happening, federal agencies are beginning to redefine their own missions to include environmental stewardship. Environmental stewardship is environmental management that goes beyond compliance and includes more proactive efforts, such as pollution prevention, natural resource conservation, and environmental management programs deemed "state of the art" by organizations, peers, and stakeholders.

Role of the Federal Facilities Enforcement Office

In the 1990s the EPA has challenged regulated entities to accept a new focus on auditing by proposing two additional areas to the audit protocol for conducting audits. The first area proposed in 1991 encouraged federal agencies to move toward beyond compliance auditing into the arena of pollution prevention. To facilitate this process, the EPA encouraged entities to conduct pollution prevention opportunity assessments in conjunction with compliance audits. At first, the focus of this guidance was on the reduction of the generation of hazardous waste, and the initiatives on this issue were typically referred to as "waste minimization" programs. However, on the passage of the Pollution Prevention Act, Congress required the EPA to encourage regulated entities to expand pollution prevention efforts to a more expanded "multimedia approach." In response the EPA developed a Pollution Prevention Opportunity Assessment Manual that provides regulated entities the procedure for auditing facilities for multimedia pollution prevention practices.

The second area proposed was auditing of management systems. In 1995 the EPA revised the 1989 version of the *Generic Environmental Audit Protocol for Federal Agencies* to include this new concept. The protocol provides guidance to auditors on evaluating management systems effectiveness of environmental programs. The latter approach provides guidance to auditors in evaluating the strengths and weaknesses of the management systems that ensures high-quality, sustained management of environmental operations throughout the entire facility or agency.

In March 1995 the EPA in partnership with auditing experts within the federal government and the Institute for Environmental Auditing held a 4-day conference on environmental auditing for federal environmental managers. The conference provided training for both experienced and inexperienced federal managers on the concept and value of environmental auditing and also provided specific training on the use of the EPA's environmental audit protocol document, *Generic Protocol for Conducting Environmental Audits of Federal Facilities*, 2nd Edition (EPA # 300-B-95-002, Feb 95).

The EPA has also recently updated the *Environmental Audit Program Design Guidelines for Federal Agencies* due for release in June 1996, which will include new concepts in audit techniques and approaches. (The previous version of this guideline was issued as EPA document No. EPA 130/4-89/001, Aug/89.)

The environmental performance expectations that have been placed on federal agencies in recent years have required the EPA's Federal Facilities Enforcement Office to focus on monitoring federal facility activities and on assisting agencies in developing and improving their compliance programs.

The relationship between the EPA and other federal agencies was first prescribed by Presidential Executive Order No. 12088 signed by President Carter in 1978. The order directs all departments and agencies of the federal government to comply with federal, state, and local laws and regulations. In addition, the EPA is directed by the order to provide technical assistance and guidance to federal agencies to assist them in complying with these environmental requirements.

To ensure that federal facilities receive the appropriate level of monitoring and guidance, the EPA established a separate office reporting directly to the Office of Enforcement and Compliance Assurance. This office is the Federal Facility Enforcement Office. The FFEO is responsible for ensuring that federal facilities take all necessary actions to prevent, control, and abate environmental pollution. The FFEO coordinates the OECA's federal facility enforcement, compliance assurance, and assistance efforts. It also has the lead role for communicating with Congress, other federal agencies, states, and other stakeholders, e.g., the public, on federal facility matters.

The design of the EPA's FFEO embodies many of the principles embraced by the EPA's Common Sense initiative. It has a sector orientation, uses strong enforcement combined with compliance assistance, and promotes proactive technical programs, such as pollution prevention and environmental auditing. The FFEO continually seeks new and innovative ways of working with federal agencies by offering technical assistance within a partnership setting with other federal agencies, states, and localities to offer a more collegial approach to environmental problem solving.

The EPA's FFEO is cochairing an interagency work group and will provide technical assistance to help federal agencies design and initiate audit programs. Although the EPA will respond to disclosure of self-audit reports in the same manner as other regulated entities, federal agencies should, however, be aware that the Freedom of Information Act will govern any disclosure of audit reports requested by the public.

Unique Aspects of Federal Environmental Audit Programs

In general, the benefits accrued by the private sector can be realized by federal government facilities. However, several important factors must be considered before designing and implementing an environmental audit program for a government agency. For example, the allocation of resources, i.e., money for compliance programs and for the hiring of sufficient and qualified staff, must be determined more by internal federal policy, e.g., Federal Acquisition Regulations, and the prerogatives of the dominant political influence affecting the agency, such as the White House staff or Congress. In addition, the process of obtaining resources for environmental programs, i.e., funds, is usually long and arduous and follows a complex process predefined by authorities outside of the organization, such as the Office of Management and Budget.

Agency Mission Versus Environmental Compliance — First and foremost, the operations of a particular federal agency are not driven by the economics of making a profit nor meeting production goals as it is in industry and the private sector. Instead, agencies have predefined missions that charter their service to the American people. The goals, objectives, and responsibilities of each federal agency are documented typically in enabling statutes written by Congress and reflected in each agency's mission statement. Therefore decisions regarding how a facility will perform environmentally and what environmental issues are of top priority are made under a more complex set of circumstances than in the private sector.

To prevent environmental problems and to stay abreast of changing legal requirements and other trends, federal agencies are beginning to redefine their own missions to include environmental management concepts that are more progressive or reflect a state-of-the-art approach. Several agencies within the Department of Defense are already developing progressive environmental programs that go beyond compliance by emphasizing environmental auditing, pollution prevention, and other proactive programs, such as the Total Quality Environmental Management concept created by the Global Environmental Management Initiative.

In the private sector proactive environmental programs, such as pollution prevention and environmental auditing, make perfect sense primarily because of business costs. Corporate officials are beginning to view environmental releases as an indication of inefficiency, i.e., failure to fully utilize resources. The private sector views this inefficiency as leading to unnecessary increases in raw materials and emission management costs that can be minimized if proactive compliance programs are employed. In addition, most corporate officials have witnessed a dramatic increase in compliance and environmental liability costs associated with third-party complaints or government lawsuits.

Despite the benefit of proactive environmental programs, the decision to "re-tool" and thus improve federal agency environmental performance is not as straight-forward as it might seem. The federal agencies are under enormous pressures to cut spending and increase staff to management ratios. The federal environmental officer is often caught in the middle, between the dwindling resources and trying to maintain an adequate environmental program. The trend today is to try and do more with less, and careful consideration must be made as to the payback on newly implemented programs and technologies. But even though the cost of investing in a better information management system for compliance programs is difficult to justify to senior agency management, the payback in cost savings associated with preventing problems is far better than dealing with the alternatives associated with sudden cost requirements, e.g., spill response, lawsuits, regulatory fines, that result from reacting to problems after they occur.

Public pressure has recently forced the federal government to reconsider how it does business. David Osborne and Ted Gaebler stated in their book *Reinventing Government*, that "anticipatory governments do two fundamental things: (1) they use an ounce of prevention rather than a pound of cure; and (2) they do everything possible to build foresight into their decision making." They also observed that "some governments are not only trying to prevent problems, they are working to anticipate the future and to give themselves radar, to prevent problems before they emerge rather than simply offering services afterward." Many organizations, both corporate and public, have now come to believe that to truly reduce liability and unnecessary costs, regulatory compliance should become the floor and not the ceiling of environmental performance. Therefore effective environmental management must balance the needs of the stakeholder in the context of inherent risks of an agency's mission, senior management's expectation for performance, and other factors, such as shrinking resources.

In all of this, the benefits of performing environmental audits are paramount to managing proactive environmental management programs. It is the performance of environmental audits and the results that will help the agency compliance officer, attorneys, and executives make informed decisions regarding environmental management and risk and the impacts of each on the agency's mission.

The manager of a federal agency environmental audit program must clearly articulate the audit program objectives for all elements of the EPA's mission, as well as its own environmental policy. To do otherwise may render the audit program a stigma of being irrelevant and unresponsive to the agency's needs. For some federal compliance officers this task can be quite complex if environmental protection is part of the agency's overall mission, e.g., Department of Interior, National Oceanic and Atmospheric Administration, or if access to environmental information affects national security concerns (military installations). Therefore these kinds of special needs should be identified in the agency's audit policy and strategies should be developed that recognize the sensitivity of these issues while accomplishing sound environmental management with a minimum of disruption.

National Security Concerns — Federal facilities with law enforcement or defense-related functions, i.e., the Department of Defense, Department of Energy, and the Central Intelligence Agency, frequently have special security requirements. Auditors must comply with security regulations for access to these facilities and associated documents. This may mean providing adequate lead time to obtain the necessary security clearances to conduct audits. Usually secure areas and documents can be protected to allow an audit to proceed without compromising national security. An internal audit program should be designed so an adequate number of auditors have the necessary clearances and so delays in audits are avoided.

As with any facility that is audited, the auditor must make a record of any areas at a facility that are not accessible for inspection and any documents or other information that cannot be reviewed due to inaccessibility. In accordance with good audit practices, the audit report should account for any areas that were not inspected or evaluated during the audit, and the facility personnel need to be made aware that the audit results are limited only to those areas that were accessible for inspection and evaluation. This prevents any false assumptions related to the scope and comprehensiveness of the audit.

Effects of the Federal Budget Cycle — The Congress and Executive Branch, including the Office of Management and Budget (OMB), Congressional House and Senate committees, are responsible for budget planning and appropriations for each federal agency. Funding for environmental management projects are typically initiated at the facility level first, usually by the facility compliance officer or person in charge of environmental management. Such projects requiring capital expenditure are typically considered "line items" in an agency's budget. Next, the operating unit or bureau of the agency must approve the funding, and, finally, the project must be approved by the agency budget official before being forwarded to the White House for approval.

An understanding of the timing involved in the budget process from one level of approval to another is critical in ensuring that the necessary funds are received in time to meet compliance deadlines or gather key data required for abatement and cleanup of contaminated sites. This concern is particularly applicable to an environmental auditing program, not only to ensure that facilities are audited in a timely manner but also to correct environmental problems found during an audit.

Another key concern is top management support for the environmental program and the understanding that facility personnel need to build compliance into their facility budgets. Senior federal executives and budget officials at the agency need to understand that environmental management has to be recognized as part of the "cost of doing business" in the government and therefore needs to be part of the agency's overall budget submission. In addition, the requests for environmental management funding should be adequately justified and prioritized, so that each reviewer in the process will understand the significance of the request.

Finally, the environmental budget request should have a distinct identity from other facility budget requests and not be "hidden" in with other types of funding categories. In this manner the request will allow the budget reviewer to understand the overall importance of the requested funding and hopefully assign an appropriate priority to the item.

Federal Agency Environmental Management Plan — Federal Agencies are required under Executive Order 12088 to submit environmental project plans to the OMB through the EPA for review.[13] The purpose of this requirement is to ensure that federal agencies identify all environmental requirements and target adequate resources to address them.

To help federal agencies comply with the executive order requirement, the EPA has offered guidance in the form of a process known as the Federal Agency Environmental Management Plan (FEDPLAN) program. FEDPLAN

---

[13] In response to Executive Order 12088, the OMB has further defined that process for federal agencies by issuing OMB Circulars A-106 and A-11.

is a reporting mechanism defined by the EPA and consists of a combination of written guidance and a PC-based desktop management information system known as FEDPLAN-PC. The guidance for using the FEDPLAN system is contained in a recently issued EPA guidance document, *Federal Agency Environmental Management Program Planning Guidance*.[14]

The FEDPLAN guidance issued by the EPA suggests that compliance officers requesting funds for environmental projects should do so by including program management costs in their environmental plans. The EPA defines program management costs as those costs required to manage environmental programs. Other personnel costs, based on estimated time dedicated to the environmental program, may be included, as well as costs for contract personnel dedicated to any media-specific compliance strategy. The program cost definition includes inventories, assessments, surveys, studies, plans, and environmental audits. The EPA proposes to make this category a subject of special analysis during the EPA reviews of individual federal facilities by the EPA regional offices. However, overall federal agency funding in this area will be reviewed and monitored by the EPA headquarters personnel.

Contractor and Tenant Activities — Certain federal environmental regulations assign responsibility for compliance by using the term "owner" or "operator" of a particular facility or site. However, identifying the owner or operator of a federal facility for the purpose of assessing responsibility for compliance and environmental management is not a simple matter. The federal agencies in many cases have developed complex relationships with other types of regulated entities, i.e., private parties, other federal agencies, and state or local governments. At many federal facilities there are tenant relationships with these types of other regulated parties and often the responsibility for environmental management and compliance is not clear between the affected parties. The following definitions describe a few of the contractor or tenant relationships encountered at federal facilities:

- Government owned/contractor operated — a facility owned by a federal agency but operated by private contractors contracted to the federal government for performing government services.
- Government owned/privately operated — a facility or lands leased by the federal Government to private operators for their own operation and profit.
- Privately owned/government operated — a facility where the government leases buildings or space from the private sector, to be operated by a government agency for government programs.

Even though the tenant may be responsible for environmental compliance of their activity, e.g., as stipulated by lease agreements, the *facility owner* may be ultimately held accountable by regulators should compliance problems

[14] Environmental Protection Agency Publication Number 300-B-95-001.

persist or should future liabilities be discovered. Therefore, before an environmental audit is performed at a joint tenant/owner or multiple tenant facility, several questions must first be answered including the following:

- Who actually owns the property and how many tenants are involved with on-site activities that may be the focus of an environmental audit?
- What is the statutory language as to who can be held responsible for on-site compliance activities, e.g., permitting, monitoring, and any abatement and cleanup of on-site contamination that may occur as a result of environmental activities?
- What are the operative instruments, including contractual arrangements specifying who is responsible for ensuring compliance with applicable environmental statutes and regulations?

The agreements between facility owner and the tenant organizations need to clearly establish environmental responsibilities of both parties, as well as the mechanisms that will be used by the responsible party to monitor compliance, including any assessments or audits previously conducted.

Freedom of Information Act — The Freedom of Information Act (FOIA), applies to all federal agencies and governs the disclosure of federal agency documents to the public. For this reason, careful consideration must be given by agency employees in determining how audit reports and audit-generated information will be filed and distributed within the agency. Generally, draft audit reports, preliminary information, and auditor notes that contain the auditor's thoughts and observations recorded during an audit site visit may be exempt from the FOIA if the information is considered predecisional by the agency's FOIA officer. Federal agencies are allowed, under the statute, to write their own policies and regulations that implement the FOIA. It is often these policies and regulations that influence the agency's FOIA officer's decision as to whether the information is releasable when it is requested by the public.

To the extent that draft copies of audit reports are predecisional and can be shown to reflect the agency's deliberative process, they may be exempt from release. However, if factual material, e.g., observations made on site during the audit, is requested under the FOIA, the agency may have to extract this material from the draft audit report and release it to the requesting party. To protect draft copies within the deliberative process, all reports and related paper should be clearly marked "predecisional, FOIA Exempt" or "draft," and circulation should be limited to those offices or audited facilities reviewing the report before producing a final version.

Legal advice from an agency's general counsel may provide additional help when processing the FOIA requests for audit-related information. In addition, the effect of the FOIA on audit-related information should be considered when designing an audit program or creating a scope of work for federal agency audits.

Conclusions

Environmental auditing at federal agencies is unique and auditors must consider several conditions and circumstances that set federal facilities apart from auditing in the private sector. However, once an auditor is aware of these differences and is able to conduct and write an audit report that accounts for these unique characteristics of federal facilities, the benefits of the product from the audit effort are the same as those experienced in the private sector.

5 PRIVATE SECTOR INITIATIVES AND POSITION

The critical issues of auditing previously discussed are important advocacy challenges to the private sector. The private sector's position is that the voluntary auditing program should be protected from disclosure and there should be an atmosphere of fair play for those taking the initiative to correct their problems. With the question of mandated or otherwise regulated audits, the industry challenge is to demonstrate that legislation is not only unnecessary but also counterproductive, because openness, candor, and flexibility will be hampered by government involvement. Industry considers that government resources should focus on incentives for voluntary compliance rather than for enforcement. Auditing is a living process that is constantly evolving to meet the needs of society and will continue to require flexibility as society's needs evolve.

The consensus of industry is that there is a need for a level playing field to create positive incentives for those taking the initiative to implement an audit program, rather than penalizing them. Disclosure risks are high and could be self-incriminating. Safe harbor protection for audit programs that have taken prompt corrective actions should provide a grace period before a violation would become subject to penalty or court action. The grace period would not only be fair, but would encourage corporations to take immediate action to audit and to then correct adverse audit findings.

Disclosure and the grace period should be closely tied together. Many companies are currently using verbal closing meetings for an audit, as well as other ways to restrict the free flow of information within the corporation. This chilling effect on voluntary audit programs is a result of the lack of statutory protection. This effect is pervasive throughout industry.

INDUSTRY SURVEY

A voluntary Environmental Audit Survey of United States businesses was conducted by Price Waterhouse, L.L.P., to develop benchmark data (Ref. 1).

The survey targeted a broad range of companies and industries. The highlights of the survey are summarized below.

Industry sectors with the highest percentage of audit use are (in decreasing order) as follows:

- Electric and gas utilities
- Aerospace or defense equipment
- Forest or paper products
- Health care products
- Chemical or petrochemicals
- Nonutility energy production or petroleum refining

The primary reasons why these industrial sectors perform audits are as follows:

- To identify problems internally and to correct them before discovery by agency inspections
- To improve the company's overall environmental program and make it proactive
- To ensure that management control systems are functioning
- To decrease the company's operating and financial risks

Ninety percent of the corporate respondents that audited did so to find and correct violations before they were found by government inspectors. Other reasons for auditing cited by respondents included lower exposure to potential third-party liability and lower insurance costs.

The factors that would encourage these industrial sectors to perform more audits are as follows:

- An enforcement policy that eliminates penalties for self-identified, reported, and corrected items
- The passage of a federal privilege law
- A presumption against company criminal prosecution if a comprehensive audit program is in place, corrections are made, and violations are reported
- An enforcement policy that eliminates or substantially reduces potential penalties for self-identified, reported, and corrected items reported within a grace period
- Customers that require that a comprehensive audit program be in place
- Passage of a state privilege laws
- The availability of *fast track* permit procedures if a comprehensive audit program is in place

The factors that discourage these industrial sectors from not expanding their audits are as follows:

- Limited company resources
- The potential for civil and criminal enforcement actions
- The possibility of citizens' suit or other litigation
- Disclosure causing a negative company image

The cost to audit a typical facility ranges from $200 to $150,000, with the mean being $15,401. The annual cost of maintaining an audit program ranges from $200 to $4,000,000, with a mean of $367,285.

The primary performance standards employed were strictly legal and regulatory compliance, as well as company policies and procedures.

The primary impact of an audit program on improving specified factors was reported to be as follows:

- Improved regulatory compliance
- Better management awareness of their responsibilities
- Reduced future environmental liability
- Improved compliance with company policies and procedures

The activities and positions of some of the private sector organizations involved in auditing issues are presented below.

INTERNATIONAL STANDARDS ORGANIZATION

The International Standards Organization (ISO) is an international voluntary consensus organization based in Geneva, Switzerland, founded in 1946 to develop international standards and reduce trade barriers resulting from national standards. The ISO is most well known for the ISO 9000 Quality Management Standards adopted in 1987. The ISO develops generic standards, and the ISO 14000 standard series for environmental management focus on management evaluation (environmental management system, environmental performance evaluation, environmental auditing) and product evaluation (life cycle assessment, environmental labeling, environmental aspects in product standards). A technical committee (TC) 207 has been set up to develop the ISO 14000 EMS. The committee is working to structure the ISO 14000 in a manner similar to the structure of ISO 9000.

Many globally competitive corporations are preparing to register under the ISO 14000 EMS standard. The ISO 14000 EMS contains specifications for auditing. These specifications must be in place and functional before registration can be achieved. ISO 14000 standards for auditing fall into three groups: auditing principles, auditing procedures, and auditor qualifications.

Further discussion of the guidance and the ramifications of the auditing requirements may be found in Chapter 9.

The EPA has submitted comments on the ISO 14000 standards. Although not all the comments were incorporated, the EPA is generally satisfied with the pollution prevention and compliance auditing language that was included. It is essential to obtain EPA's sanction of ISO 14000 to gain credibility among future registrants, because an EPA sanction would automatically lead to confidence in ISO 14000 registration by United States companies.

BRITISH STANDARDS INSTITUTE AND THE ENVIRONMENTAL AUDITORS REGISTRATION ASSOCIATION

The United Kingdom has been the leader with the development of the first international standard on EMS with the British Standard 7750 (BS7750) developed in 1992 by the British Standards Institute and revised in 1994 to reflect many of the changes from the final version of the European Community Eco-Management and Audit Scheme Regulations. Some of the most recent changes in BS7750 include the following (Ref. 8):

- A definition of continual improvement of environmental performance
- An introduction of economically viable applications of the best available technology as the environmental performance level toward which to work
- A commitment to achieve sustainable development
- A frequency of audit at least every 3 years

There are many commonalities between the BS7750, the draft ISO 14000, and other primary international EMS standards, and a comparison chart is shown in Chapter 9.

The Environmental Auditors Registration Association (EARA) is an independent, nonprofit organization based in the United Kingdom and open to international membership. The organization registers individuals for three levels of auditors.

Principal environmental auditor
Environmental auditor
Associate environmental auditor

The registration was open in November 1992 and has already received more than 1000 applications internationally. Transfer of auditors to the environmental fields from other fields such as quality, finance, safety, and health, is possible through successful completion of appropriate accredited training courses. The association is the recognized environmental register for BS7750 in the United Kingdom.

CHEMICAL MANUFACTURERS ASSOCIATION

The Chemical Manufacturers Association (CMA), the largest trade organization representing 90% of the chemical and pharmaceutical industry, advocates the Responsible Care Principles. The Responsible Care Principles indirectly prescribe an audit program by requiring members of the CMA to self-evaluate their performance annually against the ten Responsible Care Principles, the six Codes of Ethic, and the 106 Management Practices of chemical companies. The CMA's position is to dissuade government from mandating audits to maintain openness and flexibility, and to persuade government to focus its audit and enforcement policy on the encouragement of more voluntary audits and the establishment of a fair system in the competitive market.

The guiding principles of responsible care are a commitment to fully support a continuous effort in industry's responsible management of chemicals by responsiveness to public concerns. The guiding principles are as follows:

- To recognize and respond to community concerns about chemicals and industry operations
- To develop and produce chemicals that can be manufactured, transported, used, and disposed of safely
- To make health, safety, and environmental considerations a priority in our planning for all existing and new products and processes
- To promptly inform officials, employees, customers, and the public of information on chemical-related health or environmental hazards and to recommend protective measures
- To counsel customers on the safe use, transportation, and disposal of chemical products (product stewardship)
- To operate plants and facilities in a manner that protects the employee and public environment, health, and safety (ES&H)
- To extend knowledge by conducting or supporting research on the ES&H effects of our products, processes, and waste material
- To work with others to resolve problems created by past handling and disposal of hazardous substances
- To participate with government and others in creating responsible laws, regulations, and standards to safeguard the community, workplace, and environment
- To share experience and assistance with others who produce, handle, use, transport, or dispose of chemicals

To commit to and comply with the principles above, an assessment program to monitor the continuity of industry management of chemicals is compelling.

INTERNATIONAL CHAMBER OF COMMERCE AND GLOBAL ENVIRONMENTAL MANAGEMENT INITIATIVE

The International Chamber of Commerce (ICC) Business Charter for Sustainable Development contains the principles that promote openness to employees and the public, and the auditing that measures environmental performance. The purposes of the principles are as follows:

> To measure environmental performance; to conduct regular environmental audits and assessments of compliance with company requirements, legal requirements and these principles; and periodically to provide appropriate information to the Board of Directors, shareholders, employees, the authorities and the public. (Ref. 9)

The Sixteen Principles

The 16 ICC principles provide a framework for all of the major aspects of environmental management for promoting sustainable business development and are designed to improve environmental performance. The term environmental also refers to environmentally related aspects of health, safety, and product stewardship. The 16 principles, covering all aspects of environmental management, are listed below (Ref. 9).

1. *Corporate Priority* — To recognize environmental management as among the highest corporate priorities and as a key determinant to sustainable development; to establish policies, programs, and practices for conducting operations in an environmentally sound manner.
2. *Integrated Management* — To integrate these policies, programs, and practices fully into each business as an essential element of management in all its functions.
3. *Process of Improvement* — To continue to improve corporate policies, programs, and environmental performance, taking into account technical developments, scientific understanding, consumer needs, and community expectations, with legal regulations as a starting point; and to apply the same environmental criteria internationally.
4. *Employee Education* — To educate, train, and motivate employees to conduct their activities in an environmentally responsible manner.
5. *Prior Assessment* — To assess environmental impacts before starting a new activity or project and before decommissioning a facility or leaving a site.

6. *Products and Services* — To develop and provide products or services that have no undue environmental impact and are safe in their intended use, that are efficient in the consumption of energy and natural resources, and that can be recycled, reused, or disposed of safely.
7. *Customer Advice* — To advise and, where relevant, educate customers, distributors, and the public in the safe use, transportation, storage, and disposal of products provided; and to apply similar considerations to the provision of services.
8. *Facilities and Operations* — To develop, design, and operate facilities and conduct activities taking into consideration the efficient use of energy and materials, the sustainable use of renewable resources, the minimization of adverse environmental impact and waste generation, and the safe and responsible disposal of wastes.
9. *Research* — To conduct or support research on the environmental impacts of raw materials, products, processes, emissions, and wastes associated with the enterprise and on the means of minimizing such adverse impacts.
10. *Precautionary Approach* — To modify the manufacture, marketing, or use of products or services, or the conduct of activities, consistent with scientific and technical understanding, to prevent serious or irreversible environmental degradation.
11. *Contractor and Suppliers* — To promote the adoption of these principles by contractors acting on behalf of the enterprise, encouraging and, where appropriate, requiring improvements in their practices to make them consistent with those of the enterprise; and to encourage the wider adoption of these principles by suppliers.
12. *Emergency Preparedness* — To develop and maintain, where significant hazards exist, emergency preparedness plans in conjunction with the emergency services, relevant authorities, and the local community, recognizing potential transboundary impacts.
13. *Transfer of Technology* — To contribute to the transfer of environmentally sound technology and management methods throughout the industrial and public sectors.
14. *Contributing to the Common Effort* — To contribute to the development of public policy and to business, governmental, and intergovernmental programs and educational initiatives that will enhance environmental awareness and protection.
15. *Openness to Concerns* — To foster openness and dialogue with employees and the public, anticipating and responding to their concerns about the potential hazards and impacts of operations, products, wastes, or services, including those of transboundary or global significance.
16. *Compliance and Reporting* — To measure environmental performance; to conduct regular environmental audits and assessments of

compliance with company requirements, legal requirements, and these principles; and periodically to provide appropriate information to the Board of Directors, shareholders, employees, the authorities, and the public.

The ICC principles provide the framework for all aspects of environmental management; policy setting, systems and procedures, implementation and education, and monitoring and reporting. For each principle, the Global Environmental Management Initiative (GEMI) has developed environmental management system auditing elements that are presented in Table 5.1.

The GEMI is a group of leading companies dedicated to fostering business ethics worldwide for environmental excellence and sustainable development. The current members of GEMI, as of June 1995, are shown on the GEMI's Organizational Structure of ISO/TC 207: Environmental Management and United States Participation Chart, March 1995. The chart may be found in Appendix C.

The GEMI together with other associations have initiated an environmental self-assessment program (ESAP) to measure a company's performance in meeting the intentions of the 16 principles of the ICC Business Charter for Sustainable Development. The focus of the ESAP is on corporate level policy, systems, and performance measurement programs. The ESAP will allow management to prioritize and target opportunities for improvement. The ESAP is an example of an EMS audit program, as opposed to a compliance audit program.

For each of the 16 principles there are horizontal audit elements as listed in Table 5.1. The performance is measured for each audit element on four vertical performance levels in sequential order: regulatory compliance, systems development, systems integration, and total quality approach. The first vertical level, regulatory compliance, is the most rudimentary and is a measure of a company's conformance with laws and regulations. The systems development level explores corporate policies more comprehensively. In the next stage, systems integration, environmental information and issues are incorporated into the business planning functions; capital budgets; product designs, development, manufacture, and disposition (life-cycle assessment), etc.; and the direct and indirect environmental impacts of products, operations, and services. At the total quality level, the integrated environmental management systems are applied to operations globally with emphasis on continual evaluation and improvement for fostering sustainable development.

The use of the sequential order of the vertical performance levels emphasizes the importance of implementing one level throughout the company and its programs prior to moving to the next higher level, for example, integrating environmental issues within the day-to-day business operations before applying total quality. The ESAP uses a weighted scoring system for each vertical and horizontal performance level, and the scores are useful for the prioritization of management decisions. Each horizontal audit element is weighted by giving

Table 5.1 Principles for Environmental Management and Audit Elements

Business activity	Principles	Audit elements
Policy setting	1. Corporate Priority	1. Scope of Policy
		2. Management Involvement
		3. Resources
		4. Communications
		5. Implementation
		6. Accountability
	5. Prior Assessment	1. Property and Business Acquisition or Divestiture, and Joint Venture
		2. Facility or Venture Planning
		3. Site Closure Planning
		4. New Business Activity or Project Planning
	6. Products and Services	1. Environmental Impact
		2. Product and Service Safety and Integrity
		3. Energy Consumption
		4. Use of Natural Resources and Raw Materials
		5. Stewardship of Natural Resources
		6. Waste Minimization and Management
	10. Precautionary Approach	1. Process Changes
		2. Marketing Changes
		3. Changes in Products or Services
		4. Changes in Conduct of Activities

Table 5.1 Principles for Environmental Management and Audit Elements (Continued)

Business activity	Principles	Audit elements
Systems and Procedures	2. Integrated Management	1. Planning 2. Reporting 3. Information Flows 4. Control
	8. Facilities and Operations	1. Internal Operating Standards/Practices 2. Solid and Hazardous Waste Reduction and Treatment 3. Waste Residue Management and Disposal 4. Energy Minimization Program
	9. Research	1. Research on Raw Materials Procurement and Use 2. Research on Products 3. Research on Processes 4. Research on Waste Minimization and Emissions
	12. Emergency Preparedness	1. Hazard and Incident Assessment 2. Emergency Response Plans 3. Product and Service Safety 4. Employee Training
Implementation and Education	4. Employee Education	1. Awareness Programs 2. General Skills Training 3. Environmental, Health, and Safety Professionals Training 4. Management Development 5. Motivation
	7. Customary Advice	1. Advice to Customers and Distributors 2. Advice to Transporters 3. Advice to Consumers 4. Advice to the Public and to Environmental Groups

PRIVATE SECTOR INITIATIVES AND POSITION

	11. Contractors and Suppliers	1. Contractor Priority
		2. Contractor Performance
		3. Supplier Priority and Performance
	13. Transfer of Technology	1. Technology Information
		2. Management Methods
		3. Transfer to Industrial Sector
		4. Transfer to Public Sector
	14. Contributing to the Common Effort	1. Advice for Public Policy
		2. Contribution to Environmental Protection Programs
		3. Environmental Education Initiatives
Monitoring and Reporting	3. Process of Improvement	1. Technical Developments, Scientific Understanding, and External Expectations
		2. Improvement of Policies
		3. Improvement of Programs and Products
		4. Improvement of Performance
		5. Process for Change
	15. Openness to Concerns	1. Employee Workplace Concerns
		2. Customer and Consumer Concerns
		3. Community Concerns
	16. Compliance and Reporting	1. Improvement of Environmental Audits
		2. Progress Measurement
		3. Internal Performance Reporting
		4. External Performance Reporting

it an alphabetical score (A, B, or C) that indicates its importance. Each audit element is also scored numerically (1 through 4) when being compared on the vertical performance level.

The scoring should be accompanied by the documentation of the reasons for how and why a particular score was assigned. Using principle 4 and the auditing elements 4.1 and 4.2 from Table 5.1, an example of the scoring system is provided in Table 5.2.

ENVIRONMENTAL AUDITING ROUNDTABLE

The Environmental Auditing Roundtable (EAR) is the world's oldest and largest organization of professional environmental, health, and safety auditors. The EAR's mission includes promoting ES&H auditing as an important compliance tool and enhancing the profession of auditing. The EAR represents a broad spectrum of stakeholders, including manufacturing companies, audit service providers, legal and financial representatives from communities and the government, environmental interest groups, and audit software technology groups.

The EAR, founded in 1982 and currently growing at an annual rate of 33% per year, promotes ES&H auditing practices under the following four principles (Ref. 10):

- Using environmental auditing as the management control element of an EMS and a measurement method for a total quality management program.
- Current EPA and DOJ policy should be revisited to encourage voluntary audit programs.
- Mandatory audits that will limit the scope of audits to a minimum of regulatory compliance should be discouraged.
- Protection of public disclosure is critical to quality and openness of an audit program.

The EAR guideline document *1993 Standard for Performance of EH&S Audits* is now complimented by another document the *Standard for Design and Implementation of EH&S Audit Programs* (Draft October, 1995). These two documents provide excellent assistance for checking existing audit programs.

COALITION OF ENVIRONMENTALLY RESPONSIBLE ECONOMIES

The Coalition for Environmentally Responsible Economies (CERE), comprised of corporate entities, including brokers, analysts, bankers, and other investing institutes, was convened to bring about social change. The CERE,

PRIVATE SECTOR INITIATIVES AND POSITION 57

Table 5.2 Sample Scoring Sheet for Principle 4: Employee Education

Audit elements	Not applicable	Performance level				Element importance A = Greatest Importance B = Medium Importance C = Least Importance	Notes
		1	2	3	4		
4.1 Awareness Programs		Awareness program for employees directly **affected** by environmental compliance requirements.	**System** established to provide periodic **awareness programs** for most employees.	Frequent awareness programs offered to most employees. Environmental activities of company frequently discussed in internal publications or meetings. The effectiveness of the environmental awareness program is regularly **evaluated**.	General program throughout **the company** to increase the environmental awareness of all employees. Effectiveness and adequacy of awareness program are continuously evaluated **and program is updated** based on **feedback** from participants and business units.	A B C	

NA ──── 0 ────── 1 ────── 2 ────── 3 ────── 4 ── X   Score: ____

Table 5.2 Sample Scoring Sheet for Principle 4: Employee Education (Continued)

4.2 General Skills Training	Personnel are **trained in skills necessary** for compliance as part of new job and as significant changes in procedures occur.	**Formal skills training** is conducted as part of new job qualification and updated **regularly**. Scope includes technical skills and techniques for implementing corporate standards and procedures.	Formal skills training is conducted for appropriate managers and staff personnel in **all business functions**. Scope includes awareness to support goals and environmental management skills to facilitate implementation of corporate environmental policies. The effectiveness of the skills training program is **regularly evaluated**.	The effectiveness and adequacy of environmental skills training are **continuously evaluated**. Improvements are based on **feedback** from program participants and business units. Skills training system is evaluated for its contribution to achieving environmental improvements.	A B C

NA ———— 0 ———— 1 ———— 2 ———— 3 ———— 4 —— X   Score: _____

formed in 1989, has developed ten Valdez Principles based on environmental restoration, sustainable development,[15] and management commitment. The ten principles are as follows:

- Protection of the biosphere
- Sustainable use of natural resources
- Reduction and disposal of waste
- Wise use of energy
- Risk reduction
- Marketing of safe products and services
- Damage compensation
- Disclosure
- Environmental directors and managers
- Assessment and annual audit

Under this principle, audits and results would be released to the public. Not many companies have signed onto these principles, and most of industry did not play a major role in their development. The legal implications, the type and amount of information disclosure to environmentalists, etc. are the main complaints. CERE principles are directed for public involvement in changing business practice through economic pressure, different from Responsible Care Code of Practice and others that are inward-looking and that most industries favor.

CORPORATE ENVIRONMENTAL ENFORCEMENT COUNCIL

The Corporate Environmental Enforcement Council (CEEC), represented by diverse companies with strong environmental commitment and programs, focuses exclusively on civil and criminal environmental enforcement policy issues. The Council's positions on the EPA's Environmental Auditing Policy and related issues are listed below (Ref. 11):

- There is support for vigorous enforcement to punish deliberate and willful misconduct, but the enforcement is often arbitrary, inconsistent, and overly punitive.
- To be internationally competitive and achieve our national environmental goals, enforcement must be structured to promote voluntary compliance.
- Failure to offer adequate protection for audit reports creates a strong nonincentive to environmental auditing and has a chilling effect.

---

[15] Sustainable development is defined as meeting the needs of the present without compromising the ability of the future generations to meet their own needs. Sustainable development relies on the sustainable use of natural resources.

- The overriding goal of the EPA enforcement policy should be to ensure the self-evaluation programs are not creating liability.
- A qualified self-evaluation privilege for audit reports is recommended.
- Voluntary disclosure policy recommendations, not mentioned in the EPA's 1986 policy, include protection for and the establishment of criteria that reflect the importance of voluntary disclosure as a substantial mitigating factor with respect to enforcement response.
- Encourage a compliance structure where self-policing is the norm and can be achieved without massive use of government proceedings.

AMERICAN PETROLEUM INSTITUTE

The American Petroleum Institute (API) is a nationwide nonprofit trade association representing over 300 companies engaged in all aspects of the petroleum industry, including exploration, production, transportation, refining, and marketing.

The API is an active member of a broad-based ad hoc coalition of businesses and trade associations known as the Compliance Management Policy Group (CMPG). The CMPG has long advocated action by the EPA to promote more self-evaluation and other compliance assurance programs as the most effective and efficient means of attaining higher levels of compliance. The API fully supports the CMPG's 14-point statement on this issue and has observed a strong consensus on these 14 important points by the regulated community. The 14 points were published in a postmeeting statement by the API regarding the EPA's July 27–28, 1994, public meeting on its environmental auditing policy (Ref. 12). The 14 points, shared by the majority of the attendees at the July meeting, were addressed by the EPA in the *Federal Register* when the final 1995 policy was issued. The CMPG 14 points are as follows:

1. Current environmental regulations are so pervasive and complex that 100% compliance, although ideal, is not practical or achievable.
2. The EPA, as well as other governmental agency enforcement resources, is and will always be inadequate to ensure full compliance.
3. The EPA already depends largely on the regulated entities to monitor and report their own compliance. Voluntary audits with prompt corrective actions are the best and most effective ways to attain compliance and environmental goals.
4. In the private sector, corporate attitude toward compliance is more progressive than in the early days when imposition of penalties was a necessary inducement to comply, and many companies are now committed to environmental compliance and environmental leadership on their own.

PRIVATE SECTOR INITIATIVES AND POSITION 61

5. The EPA should focus its limited enforcement resources on the deliberate "bad actor" companies that choose to disregard environmental regulations rather than the "good actors" that routinely practice self-evaluation and correction in good faith efforts to comply.
6. The EPA's 1986 Environmental Auditing Enforcement Policy, restated on July 28, 1994, as the existing policy, may have been intended to promote voluntary auditing, but does not because the policy does not offer binding assurance for unwarranted use of audit information, or any protection from the imposition of penalties on companies that audit and discover areas of noncompliance.
7. Most companies believe that under the current enforcement practices, companies that discover a noncompliance problem through auditing, and promptly disclose and correct the problem, are more likely to be the target of an enforcement action than companies that do not audit, do not discover any noncompliance, and therefore do not have audit reports that may be requested by the regulators and used against them.
8. The above perception has a "chilling" effect, resulting in companies choosing either not to audit or to severely restrict the scope of the audit and thus limit the full beneficial use of audits to avoid risk of disclosure.
9. Many companies also believe that many of the enforcement agencies, including the EPA and the DOJ, routinely seek audit reports as part of their investigation leading toward prosecution.
10. Many companies still conduct audits despite the perceived risk, but try to minimize the risk by limiting the scope of the audit, restricting information (audit findings) circulation, and/or attempting to ensure attorney-client or work-product privileges at the cost of considerable time and expense.
11. These limitations, listed above, substantially lessen the usefulness of audits by the very nature of the definition of an audit.
12. The EPA needs to revise its enforcement policies in conjunction with the DOJ and the states to provide conditional assurances for good behavior and good intentions.
13. The EPA's traditional means of measuring enforcement success solely in terms of number of cases and penalties imposed should be replaced by a system that also measures success in terms of improved compliance and environmental performance.
14. State environmental audit privilege laws enacted in four states and considered by other states since the late 1993 are positive steps toward protection of audit information and should be given reasonable opportunity without undue opposition from the EPA.

The regulated community believes that the EPA's enforcement policy is an issue of considerable importance in our environmental progress and that the EPA should revise its audit enforcement and related policies to remove those inhibitions, thus encouraging more and better self-auditing and compliance assurances. The regulated entities are not demanding "rewards" for compliance. They are only asking that existing "nonincentives" to more vigorous self-evaluation be eliminated so that compliance can be improved and the community as a whole will be able to continue to make environmental progress.

The API commented on the EPA's interim policy stating that there were major flaws in the interim policy because it failed to provide the "certainty" most needed by the regulated community: the certainty that disclosure from self-evaluation to regulators will not lead to use of the audit information against the company (Ref. 13).

Although there was mention of assurance of immunity from certain civil penalties for companies meeting certain conditions, the assurances could be negated by expressly giving individual enforcement officials broad discretion to deviate from the policy in any specific case. This makes the assurances discretionary as they has always been. A policy statement cannot, by definition, be considered binding in every case. Ideally, the policy should be revised and then promulgated as a binding regulation, but as a minimum the policy should make clear the EPA expectations.

The threshold for use of audit information is not clear and too easily discourages companies from auditing. That the EPA will not seek audit reports to "trigger" an investigation does not mean audit information will not be sought and used for enforcement purposes in any given case. The reports may be sought and used whenever the EPA has "reason to believe" a violation has been committed. The policy does offer protection from criminal prosecution of a corporation, individual managers, or company employees, but does not protect them from prosecution for "criminal conduct." However, "criminal conduct" is limited to intentional violations or willful disregard of the law justifying criminal prosecution of the individuals in a corporation.

The API also commented that the "economic benefit" of noncompliance should be defined to encourage self-evaluation. At a minimum the EPA should reduce the "economic benefit" component of civil penalties by the cost of operating the self-assessment program that discovered the violation, the cost of correcting the noncompliance, and the costs incurred as a result of the noncompliance. According to the API, this offset would balance the "economic benefit" and provide for a fairer settlement. Also, consideration should be given to a company's fiscal viability and to whether actual harm was incurred by any other parties.

Finally, the API considers that the EPA response to state immunity and privilege legislation is premature and unwarranted, because there are limited experience and data to justify such a conclusion. The states should be allowed the respect and the opportunity to decide the best approach under the state law until the contrary is proven.

OTHER ORGANIZATIONS

Other organizations with additional resources include the Environmental Auditing Forum, the Institute for Environmental Auditing, Canadian Standards Associations, and the World Bank Group. Additional resources for international auditing are discussed in Chapter 11.

CONCLUSION

The seriousness attributed to the subject of auditing by the regulated community is indicated by several factors. The community paid more attention and played a more interactive role during the development of the EPA's 1995 policy than it did during the development of the 1986 policy. This is evident by the fact that there were more commenters at the public meetings and more written comments were received on the draft and final documents. The regulated community also went further by issuing press releases and statement papers. These documents stressed the need and usefulness of auditing; however, they also strongly voiced industry's concerns over regulated audits and the lack of protection from disclosure. They are seeking a level playing field and fair play from the EPA, where disclosed information will not be used for purposes of prosecution.

Another indicator of the regulated community's involvement is that many commercial institutes and private organizations are being spurred by member industries to issue guidance on auditing programs. The regulated community has embraced auditing to some extent, but is wary of the consequences and of the EPA's ability to live up to its policy.

6 LEGAL PERSPECTIVES ON ENVIRONMENTAL, HEALTH, AND SAFETY AUDITING

CONCERNS ABOUT DISCLOSURE AND PRIVILEGE[16]

For obvious reasons, members of the regulated community do not wish to spend time, money, and effort performing environmental, health, and safety self-analyses only to provide to potential opponents — the state and federal agencies, as well as possible third-party litigants — the rope for their hanging. Owners and operators of various types of properties and enterprises have therefore been extremely concerned with questions of privilege associated with the performance of environmental audits. The regulated community's concern about possible disclosure of self-audits, the several common law tools for protecting audits and their associated reports, the approach to audits government agencies have taken, and the dramatic turn in the nation's state legislatures and, potentially, in Congress toward legislated privileges for self-audits are all subjects for discussion.

Any information generated voluntarily, at significant cost, and in good faith being used as a club against the person who generated the information is troubling, especially when it is very much in the public interest that such information be developed. Audit reports are particularly sensitive for a number of reasons. First, audit reports have very obvious investigative uses. A company that hires an expert consultant to perform a compliance audit of its facility is essentially hiring someone to perform the equivalent of an agency inspection of the facility and to report back on possible deficiencies, thereby providing that company with an opportunity to recognize and correct any noncompliance. Such a report, in the hands of an agency inspector or investigator, would serve as a road map to a facility's possible areas of noncompliance. An audit performed

---

[16] The majority of this chapter, with the exception of the section on the Occupational Safety and Health Administration, was provided by Timothy A. Wilkins, Esq. (see profile preceding Chapter 1).

by a company, to help it identify and respond to environmental concerns and thereby prevent enforcement exposure, is transformed in the hands of a government agency into a document potentially capable of making the government's case against the company. In an era where knowingly violating environmental laws can result in criminal sanctions, including jail terms for responsible corporate officers, the possibility that an audit report might be used to establish not only the existence of substantive violations, but knowledge thereof, is viewed as a significant threat by companies considering an audit.

It is ironic that a company undertaking an audit to improve its environmental situation, both in terms of substantive compliance and enforcement avoidance, might legitimately see the performance of an audit as *increasing* the risk of enforcement exposure. The regulated entities are deeply offended that expensive, good faith compliance efforts might actually increase their risk of being pursued as a violator. As an evidentiary matter, a report prepared by a company's own consultant that notes potential violations will be more condemning than any evidence generated by the government. However, as a matter of public policy, self-auditing is the only realistic way to diagnose and correct environmental noncompliance, especially given the diminishing enforcement resources projected to be available to agencies in the future. As a matter of policy, government should be in the business of encouraging, not deterring, the performance of environmental audits.

Companies facing a decision on whether to perform an audit must consider the fact that, in addition to the cost of retaining a consultant and possibly attorneys to conduct the audit, their risk of liability and enforcement exposure also may rise. Competitors who choose not to audit not only avoid the cost of the audit itself, but also avoid creating evidence that might be used against them. Among the regulated community, the phrase often heard in discussions of the government's attitude toward the disclosure of audit information is, "no good deed goes unpunished."

TRADITIONAL COMMON-LAW PRIVILEGES

Under certain circumstances companies who conduct environmental audits may have some defenses to the disclosure and adverse use of audit reports. Specifically, the common-law privileges for attorney-client communications, attorney work product, and self-critical analyses may be available under certain circumstances.

Attorney-Client Privilege — In the decision of *United States vs. United Shoe Machine Corporation,* four elements were listed as required to establish the attorney-client privilege:

1. The asserted holder of the privilege is or sought to become a client.
2. The person to whom the communication was made (a) is a member of the bar of a court, or his subordinate, and (b) in connection with this communication is acting as a lawyer.

3. The communication relates to a fact of which the attorney was informed (a) by his client, (b) without the presence of strangers, (c) for the purpose of securing primarily either (i) an opinion on law or (ii) legal services or (iii) assistance in some legal proceeding, and not (d) for the purpose of committing a crime or tort.
4. The privilege has been (a) claimed and (b) not waived by client."[17] *United States vs. United Shoe Machine Corporation,* 89 Federal Supplement 357, 358-59 (D. Massachusetts 1950).

Companies commonly use the attorney-client privilege in environmental self-investigations to protect them from potential adverse disclosure of not only the lawyer's communications regarding the client company's compliance status, but also the information developed in an audit that is designed to provide attorneys with technical and factual assistance in their provision of environmental legal counseling. The attorney-client privilege is designed to allow attorneys to freely counsel their clients without fear that the counsel will be disclosed, and to encourage clients to confide in their attorneys. This allows the attorneys to provide sound advice and ensure a fair and competent defense. This rationale has been accepted in the context of environmental audits conducted to provide attorneys with technical information needed to assess compliance with environmental laws and regulations. (See *Olen Properties vs. Sheldahl, Inc.* in the Major Case Histories section of this chapter.) As a practical matter, environmental audits structured to serve this purpose only will pass muster in a claim of attorney-client privilege.

Notably, however, some courts have held that audits conducted through in-house counsel may, under some circumstances be for the purpose of providing business advice rather than legal counsel and thus are ineligible for the attorney-client privilege.[18] The appearance that the audit is part of a company's routine also is considered to reflect a dominance of business rather than legal concerns.[19]

Attorney–Work Product Privilege — The attorney–work product doctrine, originating in the seminal Supreme Court decision in *Hickman vs. Taylor*,[20] held that documents containing the mental impressions of attorneys

---

[17] *United States vs. United Shoe Machine Corporation,* 89 Federal Supplement 357, 358–59 (D. Massachusetts 1950).

[18] *United States vs. Chevron U.S.A., Inc.,* 1989 U.S. District Court LEXIS 12267, at *15 (Eastern District Pennsylvania October 16, 1989).

[19] Terrell E. Hunt and Timothy A. Wilkins, Environmental Audits and Enforcement Policy, *Harvard Environmental Law Review,* 16 p. 365, 379 (1992).

[20] *Hickman vs. Taylor* 329 U.S. 495 (1946).

need to be protected from disclosure if attorneys are to give properly searching and candid legal advice. Documents throughout which the lawyer's mental impressions are intermingled are, however, the only ones eligible for this protection. The criteria to protect audit results under the work product doctrine are as follows:

1. The audit must have been prepared in anticipation of litigation.
2. The (attorney–work product) privilege must not have been waived by a breach of confidentiality.
3. The materials in question must not have been developed to further an unlawful end.[21]

The application of the work product doctrine is more problematic than the application of attorney-client privilege when applied to the protection of audit results. The raw technical data of an audit result provided to a lawyer is obviously not interwoven with the lawyer's mental impressions. In addition, only audits conducted in anticipation of imminent litigation and in preparation for the defense of that litigation will be treated as subject to the doctrine.[22] Therefore the more socially valuable, proactive, compliance audits and other routine business, environmental audit–related practices may not be eligible for protection under this doctrine, while evaluation legal documents prepared in response to threatened litigation are likely to be protected.

Self-Critical Analysis Privilege — Some courts provide a common-law privilege to prevent disclosure of internal self-criticism that is in the public interest. The privilege originated in the 1970 case, *Bredice vs. Doctor's Hospital*.[23] In this case a plaintiff attempted to obtain the records of the hospital's internal physicians' review committee, which had considered and attempted to respond to issues raised by an accident. The accident under analysis by the committee was the subject of the lawsuit, and the plaintiff argued that the committee's records were a qualified voluntary self-policing and self-disclosure action subject to discovery.

In its decision the court held that an interest in candid analysis and action by the physicians under such circumstances outweighed the need of the plaintiff to discover the records, and found that the records were privileged as a matter of federal common law.

In the environmental auditing context, it seems obvious that such a privilege should also logically apply. However, the courts in *United States vs.*

---

[21] Id. at 384.

[22] *In re: Grand Jury Investigation,* 599 F. 2d 1224, 1229 (3rd Circuit 1979); *SEC vs. Worldwide Coin Investors,* 92 F.R.D. 65, 66 (Northern District Court Georgia, 1981).

[23] 50 F.R.D. 249 (D.D.C. 1970).

Dexter Corp.[24] and *Reichhold Chemicals, Inc. vs. Textron, Inc.*[25] have decided differently on whether a self-critical analysis privilege is available for internal corporate environmental self-criticism. This privilege was denied in the former case and upheld in the latter. While the argument is readily available to companies, no company adverse to risk should rely on this privilege under existing law alone to protect an audit from discovery.

FEDERAL GOVERNMENT RESPONSE

The federal agencies have reacted strongly to discussions of protection for environmental audits, focusing on possible losses to their investigative and enforcement capacity if access to audits were restricted. Nonetheless, the agencies continue to encourage private environmental auditing efforts.

In the 1980s when confronted with arguments that use the of corporate environmental self-audits results during investigative and enforcement activities deterred audits, the EPA attempted to set the record straight. The EPA published its nonbinding auditing policy statement in 1986 in which it stated that it "would not routinely seek" audit reports during its investigations and that it would give preferable enforcement treatment to companies that audited, disclosed audit results, and promptly corrected any discovered compliance problems.

In the ensuing years, well-publicized incidents involving harsh punishment against Coors and Weyerhauser of violations that were discovered through voluntary self-auditing, promptly corrected, and timely reported to the agencies resulted in a prevailing cynicism among members of the regulated community about the EPA's policy of "not routinely seeking" audit documents. By the early 1990s, complaints about the EPA's audit policy had reached a fever pitch.

In 1991 the DOJ published a guidance document that added fuel to the fire. In attempting to encourage auditing, the DOJ's policy stated that it would look favorably in determining whether to file criminal charges on companies that conducted environmental audits and voluntarily self-disclosed discovered violations to the relevant agencies.[26] This policy left the back-handed implication that companies who audited and, in attempting to preserve attorney-client or other privilege, did not disclose the results would face an increased likelihood of criminal prosecution. Rather than encouraging auditing, the federal government had potential auditors running scared.

---

[24] 132 F.R.D. 8 (D. Connecticut, 1990).

[25] 157 F.R.D. 522 (Northern District Court Florida, 1994).

[26] Factors in Decisions on Criminal Prosecutions for Environmental Violations in the Context of Significant Voluntary Compliance or Disclosure Efforts by Violator, the Department of Justice, June 3, 1991.

In the face of this overwhelming opposition to its existing audit policy, the EPA held lengthy public comment sessions beginning in 1994 to develop a new policy toward environmental audits and the companies that conduct them.[27] While again refusing to bind itself to an enforceable commitment, the EPA nonetheless pledged not to use environmental audits as an investigative tool. Perhaps more importantly, the EPA offered definite incentives for disclosure of audit reports, providing that it would not refer for criminal prosecution any violation discovered in the course of a voluntary self-audit and disclosed to the relevant agencies under specified procedures and circumstances. Further, it offered to eliminate the gravity-based component (generally, the largest share) of any cash penalty assessed against a company that had performed a self-audit and disclosed a discovered violation. The regulated entities, feeling that the agency had not gone far enough, criticized the EPA's refusal to bind itself and the burdensome procedures required for self-reporting to qualify as voluntary self-disclosure.

In December 1995 the EPA issued its final policy, which largely reaffirmed the posture found in its earlier interim policy.[28] Although the 1995 policy contains a new provision that includes incentives for performing audits, the policy insists that disclosures of a discovered violation must occur within ten days to be considered sufficiently prompt. The EPA has taken a significant step toward encouraging audits by providing limited penalties and rejecting criminal referrals on violations discovered by the regulated entities during an environmental compliance evaluation and promptly disclosed to the relevant agencies. However, the EPA has also left limits on the policy, which make it nonbinding, and has provided a complex eight-part list of conditions on discovery, disclosure, and company conduct, which make it difficult to be sure that a given audit and disclosure is covered by the policy. In addition, absent the enacting of federal privilege legislation, it has become clear that audit privilege or other protection from third-party litigants will not be available at the federal level. While the EPA's final audit policy offers valuable new incentives for auditing, these and other remaining areas of uncertainty leave some significant existing nonincentives in place. The final policy still leaves the fundamental nature of the interim policy unchanged.

THE STATE LEGISLATORS RESPOND

In response to the perceived failure of the EPA to convincingly limit its access to environmental audits or to reward companies who conduct environmental audits, a number of state legislatures, 14 at latest count, have moved

---

[27] Voluntary Environmental Self-Policing and Self-Disclosure Interim Policy Statement, 60 *Federal Register* 16,875 (Apr. 3, 1995).

[28] 60 *Federal Register* 66,706 (December 22, 1995).

in the last three years to adopt privilege and immunity provisions for voluntary environmental audits.[29] Such state legislation generally creates a privilege in any administrative, civil, or criminal proceeding for any report generated during a voluntary environmental audit unless it is otherwise required to be reported to a relevant agency. Exceptions to the privilege are provided for violations discovered through such an audit, but not promptly corrected by the company. In addition, some states provide immunity from any administrative, civil, or criminal penalty for violations discovered through a voluntary audit, diligently and timely corrected, and promptly disclosed to the relevant agencies. These provisions, while ensuring that discovered violations are promptly corrected, also offer some comfort to companies facing the difficult decision of whether to audit. Importantly, bills have been presented in the past session of Congress in an effort to create a comparable privilege and immunity statute applicable at the federal level. With the current composition of the Congress and present efforts to reform environmental law and the administrative state, passage of such a bill in the next two years seems unlikely.

MAJOR CASE HISTORIES

Despite the extraordinary degree of controversy over privilege issues with respect to environmental audits, there has been little published case law directly on point. However, there have been several original and influential cases, described below, involving the privilege issue.

Olen Properties vs. Sheldahl, Inc., No. CV 91-6446-WDK(Mcx) (Central District of California, April 12, 1994). In *Olen Properties* an individual had prepared an "environmental audit memorandum" for the purpose of assisting one of the defendants' attorneys "in evaluating compliance with relevant laws and regulations." On a motion to compel disclosure of the memorandum, the federal district court held that under the circumstances, it appeared that the report had been "prepared for the purpose of securing an opinion of law" and that they were reasonably "necessary to obtain informed legal advice." On these facts the audit memorandum was therefore "privileged and need not be produced." *Olen Properties* is a brief decision that provides an apparently unconditional precedent that environmental audits conducted pursuant to the *United Shoe* test cited above, will be treated as attorney-client privileged.

United States vs. Chevron U.S.A. Inc., 1989 United States District Court LEXIS 12267 (Eastern District of Pennsylvania, Oct. 16, 1989). In *Chevron,* an environmental audit was commissioned by in-house counsel which appeared to otherwise agree with the traditional requirements of attorney-client privilege doctrine. However, in this case the federal district court viewed the fact that the audit was conducted by in-house counsel, a company employee with an arguably nonlegal purpose for conducting the audit, as relevant to

---

[29] *State Privilege Legislation Multiplies in 1995; Predictions Differ About 1996,* Daily Environment Report, Bureau of National Affairs, August 29, 1995, at AA-1 to AA-4.

satisfaction of traditional tests for the applicability of the attorney-client privilege. As Chevron, the proponent of the privilege, failed to satisfy the court that in-house counsel conducted the audit for a predominantly legal, rather than business, purpose, the court found the privilege inapplicable. This view was echoed in *Ohio vs. CECOS International, Inc.,* No. 85-CR-5240C through 85-CR-5263C (Ohio C.P. April 23, 1986). The court considered that the direction by General Counsel of an audit of environmental management did not render the audit results eligible for privilege because it had predominant business/management purpose. While *Olen Properties* lays out a strong case that environmental audits can be structured to take advantage of the attorney-client privilege, the *Chevron* and *CECOS* precedents suggest that caution must be exercised in documenting a strictly legal purpose for the audit, especially when in-house counsel is involved in the audit.

Reichhold Chemicals, Inc. vs. Textron, Inc., 157 F.R.D. 522 (Northern District Court Florida, 1994). In *Reichhold,* a federal district court recognized for the first time, in an environmental auditing context, the applicability of the common law self-critical analysis privilege. On the basis of the public interest in environmental auditing, in "candidly assess[ing] . . . compliance with regulatory and legal requirements," the court found that a shield should be created to preclude the current owner of a contaminated industrial site from using an audit report against the former owner of the site in a suit under CERCLA. However, it must be noted that the decision of *Reichhold* is expressly limited to audits that retrospectively investigate a spill, accident, or other such event. While the court does not rule out a similar privilege for prospective compliance auditing, the case is only valuable as precedent by analogy for companies seeking protection under the self-critical analysis privilege.

United States vs. Dexter Corp., 132 F.R.D. 8 (D. Conn. 1990). Faced with a claim of self-analysis privilege for audits of compliance performed by a defendant in a Clean Water Act enforcement action, the *Dexter* court held that while there was such a privilege based on a finding of "public interest," in the Clean Water Act context the Congress had "intimated its will, however indirectly," that polluters face strict enforcement. As enforcement was in the public interest and protecting the audit reports might impair the instant enforcement effort, the court concluded, no self-criticism privilege applied. The *Reichhold* precedent, combined with general trends on the issue of privilege for environmental audits, challenges the extremely narrow view of the "public interest" taken by the *Dexter* court. For the time being, however, *Dexter* is a continuing precedent that should be considered by any company that seeks to rely on the self-critical analysis privilege as a sufficient means of protecting sensitive audit documents.

ENFORCEMENT STATISTICS

The aforementioned legal and policy battles over privileges covering environmental audits have taken place in the context of a general upswing in environmental enforcement, both at the state and federal levels. General trends in EPA enforcement in recent years have shown increases in the number of enforcement actions and in the severity of penalties that result from those actions. Comparable developments have occurred in state enforcement in recent years as well.

There are several reasons for these developments. Most importantly, in federal and state legislation, stiffer civil and criminal penalties have been provided in the last decade for environmental offenders. Civil penalties of up to $25,000 per day per violation are authorized under most of the federal environmental statutes, and in providing for criminal sanctions the new federal laws authorize substantial prison terms. Federal sentencing guidelines now in place render mandatory the exercise of the statutory authority to incarcerate certain environmental offenders. As a result, real prison terms without parole have become a realistic possibility in the federal system for knowing violators of the environmental laws. Plainly, the authority for a sterner approach to environmental enforcement has been provided.

In general, the EPA has accepted the statutory challenge with regard to stepped-up environmental enforcement efforts. Each new year generally brings a record high in such statistics as numbers of civil and other judicial cases initiated, numbers of referrals to the Department of Justice, and quantities of penalties assessed, both in the numbers of years of incarceration and in the amount of fines. In 1994 the agency brought a record total of 2246 enforcement actions with sanctions. Total civil penalties and criminal fines rose from $66.8 million in fiscal year 1990 to $151 million in fiscal 1994. The Office of Criminal Enforcement initiated 525 cases in fiscal 1994, a drastic increase over the 112 cases initiated in fiscal 1990. Against this backdrop the regulated community has expressed proper concern about providing evidence against themselves by conducting good faith environmental audits.[30]

Notably, the number of civil and other judicial enforcement actions by the EPA appears to have declined sharply in fiscal year 1995.[31] These reductions, however, are most likely attributable to a refocusing by the President and the EPA on pursuing the most serious and complex enforcement issues and to a shift toward a procompliance rather than enforcement-based approach on less significant violations.[32] The EPA also has been very clear that the sharp upward trend in civil and other judicial actions filed by or on behalf of the agency over the previous decade is not over. The EPA's new direction increases its selectivity, but the agency remains committed to pursuing tough enforce-

---

[30] *Environmental Protection Agency, Enforcement and Compliance Assurance Accomplishments Report,* Fiscal Year 1994 (May 1995, 4-1 through 4-3).

[31] *Environmental Protection Agency, Officials Fear Declining Enforcement Numbers Underestimate Success,* Inside Environmental Protection Agency, Weekly Report at 1, 10 (Oct. 6, 1995).

[32] *Environmental Protection Agency, Officials Attribute Drop in Enforcement to New Compliance Policies,* Inside Environmental Protection Agency, Weekly Report 16 (November 3, 1995).

ment. In other words, a real and substantial federal environmental enforcement threat to the regulated community will continue into the foreseeable future.

Comparable legislative changes and agency activity pointing toward serious, stepped-up environmental enforcement have occurred at the state level as well. It is in this context that companies considering environmental audits have been forced to evaluate whether the performance of such an audit will increase or reduce their potential enforcement exposure. In light of the mere potential for increased enforcement exposure associated with conducting an environmental audit, companies do not audit as frequently or candidly as they might, and many choose not to audit at all.

OCCUPATIONAL SAFETY AND HEALTH ADMINISTRATION

It is becoming necessary for anyone involved in environmental services to be familiar with the Occupational Safety and Health Act and for those in the safety and health field to have familiarity with the major environmental laws and regulations. Workplace contamination usually implies environmental releases. The overlap of these two disciplines is evident when parallel results are obtained from the performance of environmental auditing and from health and safety auditing.

The Occupational Safety and Health Administration's (OSHA) Hazard Communication Standard in part serves as the basis for defining the scope of an employer's obligation under the Emergency Planning and Community Right-to-Know Act. The duplication of the process safety management requirements under the Clean Air Act Amendments for both OSHA's 1992 standard and the EPA regulations indicated the inseparability of the relationship of workplace media from other environmental media. Various other federal statutes provide the EPA with authority over certain worker activities. Examples include the worker protection standards under the Toxic Substances Control Act for asbestos in schools and from numerous new chemical substances and under the Federal Insecticide, Fungicide, and Rodenticide Act for agricultural field workers. These are some of the examples of how the OSHA regulations are intertwined with the environmental regulations (Ref. 17).

Enforcement History

In the mid 1980s there was a perceived need for stiffer OSHA penalties as relatively few safety and health cases have ever resulted in criminal prosecution. Between 1982 and 1993 the Department of Labor referred only 93 cases to the DOJ for criminal prosecution, of which 52 cases were declined. The *OSHA Federal Standards Citations*[33] statistical summary for fiscal year 1994 showed that the most frequent violations were under 29 Code of Federal Regulations (CFR) 1910.1200, Hazard Communications, and the highest initial

---

[33] These are federal statistics and do not include any state occupational safety and health data.

penalty levied was for $4.8 million under the violation of the General Duty Clause (Ref. 14).

In 1986 OSHA adopted its Egregious Penalty Policy, which greatly magnified its enforcement actions through a system of a separate citation and fine for each instance of a violation. Previously, OSHA grouped similar violations and applied one penalty to that group. Under the Egregious Penalties Policy, OSHA began issuing penalties for each violation, even if violations were similar. This instance-by-instance policy applies to violations of regulations concerning the keeping of records, safety and health standards, and the General Duty Clause of the Act. Thus, if five machines had safety guards missing, each instance would be cited as a separate violation with a separate penalty, instead of grouping these instances as one violation and issuing a single citation and penalty.

Furthermore, when the penalty structure was changed by the 1990 Budget Reconciliation Act, the maximum penalty limit increased sevenfold, creating a dramatic increase of the total amount. In later cases OSHA took the position that each employee exposed to a particular hazard constituted a separate violation and when they began citing employers with this additional multiplier, even small employers were faced with large fines exceeding $1,000,000 (Ref. 15).

The *Hartford Roofing* case is an example of this multiplier in effect. Six roofers were working on a roof without the proper fall protection system, and because each of the six were exposed to fall hazard, six citations of $35,000 each were issued. A total penalty of $210,000 could have been avoided by a single act of installing a guardrail that could have protected all six workers (Ref. 16).

In another case there was an explosion and fire in Arcadian's Lake Charles, Louisiana, fertilizer plant. OSHA alleged a violation of the General Duty Clause, which requires that employers provide "each of his employees . . . a place of employment free from recognized hazards," and proposed a penalty of $50,000 for each of the 87 employees at the plant for a total $4,350,000 fine.

When the administrative law judges rejected OSHA's employee-by-employee multiplier method, both cases were appealed to the Occupational Safety and Health Review Commission. However, the decision of the Commission this time declined to follow its earlier decision in the *Caterpillar* case, involving failure to record individual cases of recordable illnesses or injuries. In the *Caterpillar* case the commission stated that each instance of a failure to record was a separate act, and the regulations require employers to "enter each recordable injury or illness on the log and summary." In contrast, the *Hartford* and *Arcadian* cases only required the employer to address the workplace hazard by implementing a single course of action, and once that hazard was addressed, all affected employees would be protected. The distinction between these decisions may seem at first difficult to comprehend. However, by following the example of the OSHA respirator standard that the commission used to clarify the distinction, the differences become understandable. Under the standard, the failure to provide each employee with respiratory protection in a hazardous air environment represents a separate and distinct violation for each. However, if the solution to the hazardous air is a single course of action, for example to provide engineered ventilation control, then only a single violation exists. Although OSHA can use the exposed employees multiplier to emphasize the greater penalty for the violation, it may not increase the number of citations for each exposed worker when a single abatement practice, method, or condition would protect more than one employee (Ref. 16).

In summary, OSHA's recent enforcement strategy of the Egregious Penalties Policy program includes the following (Ref. 16):

- An increase in the number of instance-by-instance violation cases under its egregious case policy
- Expanded use of OSHA's existing penalty authority
- More criminal referrals to the DOJ when appropriate
- Reinvention of the inspection process to cover more workplaces through better targeting and streamlined inspection procedures

Voluntary Protection Program

OSHA established its Voluntary Protection Program (VPP) in July of 1982. It was designed to augment OSHA's enforcement efforts through encouragement and recognition of excellence in voluntary occupational safety and health initiatives. The VPP was the genesis for the EPA's Environmental Leadership Program. VPP recognition is for those companies that can demonstrate a commitment to workplace safety and health beyond the regulatory requirements, especially at senior management levels. There are three levels of recognition and participation — Merit, Star, and Demonstration Programs. Companies participating in Merit and Star Programs are expected to have a program that includes employee participation and annual comprehensive self-evaluation elements. The benefits are public recognition of excellence and exemption of the work site from OSHA's programmed inspections. The VPP is expected to receive more attention by OSHA in light of the recent federal budget reductions.

The VPP is open to any industry, and the general requirements for qualification include the following:

- An effective, ongoing safety and health program — A strong program is the central element that qualifies a company. OSHA assesses the effectiveness of the program through a number of measures including on-site reviews.
- A cooperative atmosphere — A cooperative atmosphere is essential to making a voluntary protection program work and to encouraging employee participation.

- Good performance — Performance is monitored to ensure that the efforts are in fact working to minimize injury and illness in the workplace. Two indicators used are the Bureau of Labor Statistics injury incidence and lost workday injury rates.

The Star Program is targeted for companies with comprehensive, successful safety and health programs as demonstrated by their three-year average incidence and lost workday case rate being at or below the national average for their industry. They company must also meet the requirements of extensive safety management systems. The incident rates are reviewed annually, but the participant's program is evaluated every three years.

The Merit Program, an effective stepping stone to the Star Program, uses more general management systems. The participants must agree to specific goals for reducing incident rates to a level below the national average for their industry. The company is evaluated on-site annually under this program.

The Demonstration Program provides a basis for investigating promising, alternative safety and health program approaches to allow for special industry operations, such as logging, maritime, etc. Successful alternative approaches developed under the Demonstration Program will be considered for inclusion in the Star Program.

OSHA's initial application review includes a thorough review of the participant's safety and health program, on-site examination of records and logs, a review of the internal inspection history, an assessment of site conditions, and interviews of management and employees. Annual evaluations of the Merit and Demonstration Program participants and three-year evaluations of the Star Program participants include comparisons of the company's injury and illness rates with the average national industry rates to ensure that the companies continue to meet the requirements of the VPP. Additionally, at Merit Program sites, progress made toward the Star Program requirements is measured. Despite qualification in such programs, OSHA retains the responsibility for inspections in response to formal, valid employee complaints, significant chemical leaks and spills, and workplace fatalities and catastrophes.

VPP participants have observed the following benefits from participation in these programs:

- Improved employee motivation to work safely, leading to better quality and productivity
- Reduced workers' compensation costs
- Recognition in the community
- Improvement of programs that are already good, through the internal and external review that is part of the VPP application process
- That the VPP sites generally experience from 60 to 80% fewer lost workday injuries than would be expected of an "average site" of the same size in their industry

In a cooperative atmosphere between industry and an enforcement agency, a comprehensive and effective auditing program is further demonstration of industry's acceptance and readiness to participate in the VPP. A corporate auditing program for safety and health makes a facility more readily acceptable to the VPP program and gives the company favorable public recognition in addition to overall performance improvements.

OSHA's VPP and the EPA's ELP are the enforcement agencies' efforts to reach out, assist, and collaborate with industry and other participating facilities in a cooperative approach. In light of these programs, a corporate audit program is an opportunity for industry to demonstrate their initiative to measure and achieve ES&H performance goals. The VPP has been around longer than the ELP, but OSHA's enforcement and criminal prosecution cases are overshadowed by environmental enforcement and criminal prosecution cases. This is because of the larger penalties, the more intense media coverage, and simply the larger number of reported environmental penalties.

The 1996 fiscal year $312 million budget for OSHA was reduced by $16 million halfway into the fiscal year. Because of this, the effect of the reduction was doubled. Further reductions were proposed in fiscal year 1997. Similar to the EPA, the Department of Labor will probably face more budget pressure for the foreseeable future. The effect of the 1996 reduction was the evolution of OSHA's inspection and enforcement toward a more focused program. The focused enforcement activities will pose a greater risk to those in the industry sectors that are not part of the VPP. As the budget battle continues, OSHA will be using its resources to focus on the areas that will produce the greatest enforcement results.

State Safety Programs

In addition to the federal program, there are state programs. Under the OSH Act a state may assume responsibility for the development and enforcement of the OSHA standards. The states promulgate and enforce their own standards, which are subject to approval by the Secretary of Labor. In most cases the states have simply adopted federal standards, but some states have enacted more stringent standards. Employees are covered in 23 states and two territories by plans that have been approved by OSHA. The states with approved plans are as follows (Ref. 17):

Alaska	Arizona	California	Connecticut
Hawaii	Indiana	Iowa	Kentucky
Maryland	Michigan	Minnesota	Nevada
New Mexico	New York	North Carolina	Oregon
South Carolina	Tennessee	Utah	Vermont
Virginia	Washington	Wyoming	

The territories with approved plans are Puerto Rico and the Virgin Islands.

How a national company responds to the different auditing and disclosure laws within the states where they have facilities is a question yet to be answered. For instance, if there is protection from disclosure in state A and the company discloses its audit results, what is to prevent state B, where there is no legal protection, from obtaining the audit results from state A, possibly under a freedom of information request? Even if the audit results disclosed in state A are facility specific, they still could be used as a road map for inspection at the company's facility in state B. There is not enough information or precedence at this time to answer the question. It is strongly recommended that the this subject be thoroughly discussed with your legal counsel.

There are nonvoluntary programs, such as the Maine 200 Program. OSHA's Maine area office initiated a program in 1994 to improve the state's high injury rates and the corresponding worker's compensation. Maine obtained from the state worker's compensation board a list of the top 200 employers with the highest potential for improvement. These employees were given a choice to either participate in the program or be placed on a primary inspection list. Employers on the list would be subject to wall-to-wall inspections and full enforcement practices. Participation in the program meant agreement to do the following:

- Conduct a survey and inventory all hazards
- Prepare a plan for correcting hazards identified
- Prepare a written health and safety plan
- Agree not to require a warrant for an OSHA inspection
- Submit quarterly reports to the area office on progress toward completion of the program

OSHA's formal study of Maine's program shows that a majority of employers had improved results as measured by a variety of injury and illness statistics. However, a significant minority did not, and it remains to be seen if the effectiveness of the program will persist. Presently, there are plans to extend this program to other states, both at the state plan level and in federal enforcement states. About a dozen states currently have such programs.

Proposed Reform

The new administration is focusing on reevaluating the regulatory process. OSHA has been criticized for not being able to produce standards on a large number of hazards in use in the workplace and for not successfully updating their original list of chemicals and the associated Permissible Exposure Limits that was invalidated by the 11th Circuit Court of Appeals in 1992.

In 1992 Senate Bill S 575, introduced by Senators Edward Kennedy (Democrat—Massachusetts) and Howard Metzenbaum (Democrat—Ohio), and a similar House of Representatives bill, HR 1280, introduced by Congressman

Bill Ford (Democrat—Michigan), are the first major reforms of the Occupational Safety and Health Act (OSH Act) in 24 years. The bills, responding to OSHA's limited budget and resources, sought employer-employee cooperation for more effective protection of the worker. They were aimed at reducing the estimated 10,000 job-related deaths and 1.7 million disabling injuries. Labor Secretary Robert Reich, after reviewing the data on workplace injury and illness concluded that the "status quo is simply unacceptable," and that the act required improvement with an emphasis needing to be placed on employer-employee cooperation and involvement (Ref. 18).

The Democrat OSHA reform legislation proposed the following:

Reform in the Permissible Exposure Limits Setting Process — The National Institute for Occupational Safety and Health created by the Occupational Safety and Health Act, but placed within the Department of Health and Human Services, conducts research into the development of new health and safety standards, which are presented to the Secretary of Labor for consideration. Under the reform the National Institute would have to periodically (at least once every three years) submit recommendations for revisions of the Permissible Exposure Limits to the Secretary. OSHA would then publish the proposed recommendation in the *Federal Register* within six months and the final rule changes within a year of the National Institute's recommendation. Secondly, OSHA will have to revise the Permissible Exposure Limits for the list of 428 toxic chemicals that was published in the *Federal Register* in 1989 (54 FR 2332), but invalidated by the United States Court of Appeals in 1992. Third, OSHA is to promulgate similar standards for air contaminants for the construction, agriculture, and maritime industries based on the proposed rule published in the *Federal Register* (57 FR 26001).

Exposure Monitoring and Health Surveillance — Exposure monitoring standards must be promulgated within two years from the effective date of the proposal and must include a formal exposure assessment strategy, regular monitoring and measuring for chemical and physical stressors, and provide written employee notification of the exposure above the Labor Department standards. The health surveillance program must include employee exposure evaluation to determine vulnerability to risk and the periodic medical evaluation of those at risk. The employee is to be notified of the results of the medical evaluation and is not to suffer from discrimination because of the results of the evaluation.

Enforcement — Because of the perception that enforcement has not been targeted at the areas of greatest risk, the reform bills propose a special emphasis inspection program and a publication of a list of the high-risk industries and operations that will be covered in the special emphasis program. The period of abatement of a hazardous condition is to begin on the date of issuance of a citation, and if the condition is an imminent danger, the employer is to take immediate corrective action or be fined $10,000 to $50,000 per day. Finally, the maximum criminal penalty for willful violations is increased to not more than 5 years.

Employer and Employee Participation — There is greater emphasis on giving the employees a voice in the safety and health programs. Employers with greater than 11 employees are required to establish "safety and health committees" with equal representation of employees and employers who would be authorized to review the safety and health program, conduct inspections, and make recommendations.

Applicability of the Occupational Safety and Health Act — The OSH Act states that "Nothing in this chapter shall apply to working conditions of employees with respect to which other federal agencies . . . exercise statutory authority to prescribe or enforce standards affecting occupational safety and health." Thereby, the United States Department of Transportation was responsible for OSHA compliance in the operation of common carriers, including vehicles and pipelines. The reform bill attempts to alter the relationship between OSHA and other federal agencies. The review and approval of the Secretary of Labor would be required of all standards of alternative agencies to ensure that they are at least as stringent and effective as those under OSHA enforcement. If the secretary concurs with the alternative agency's standard, then the secretary must publish in the *Federal Register* a notice that jurisdiction has been ceded to that agency. This secretarial review acts as an umbrella approval authority over the OSH standards of every federal agency and would reverse the long-held congressional position that duplication of regulatory effort is to be avoided. Note, however, that this secretarial review does not apply to mining safety.

The Democrat reform bill was disputed in committee, and when the new Republican majority arrived in 1994, the approach to reform radically changed. Senator Nancy L. Kassebaum (Republican—Kansas) has introduced a Republican version of an OSHA reform bill in the Senate, while Cass Ballenger (Republican—North Carolina) has introduced his own version of an OSHA reform bill in the House of Representatives. The proposed Senate bill would exempt certain businesses from OSHA inspections and would require employees to report safety violations to their employer first, before they are allowed to report them to OSHA. Organized labor wants these two provisions struck from the Senate Bill and the White House concurs.

The House bill has provisions that require regulatory risk assessment and cost-benefit analysis that organized labor and the administration opposes. In any event, it is unlikely that there will be an OSHA reform bill passed this year, but the current activity will be the platform for the issue to be debated in 1997 (Ref. 19).

Auditing Policy

OSHA conducts scheduled and prompted inspections. However, OSHA does not have a policy similar to the EPA's policy on self-evaluation and disclosure. OSHA requires all employers to have a safety and health program that ensures workplace safety, and OSHA encourages self-evaluation of the program through its VPP.

When OSHA observes major violations during an inspection, it will routinely ask for the internal review reports. If the company does not respond to OSHA's request, then OSHA will seek the issuance of a subpoena for the records. OSHA's record in obtaining subpoenas has been quite successful for two reasons.[34] The threshold criteria for issuance of a subpoenas are low, and the OSH Act indicates OSHA's clear authority to access internal records.

The disclosure has been less of an issue in OSHA cases than it has in the EPA cases. The release of internal reports was not contested in the past, but this practice is changing. This is because the responsibility for correction of noncompliance issues is being elevated to higher management the same way environmental auditing is shifting toward management system auditing. The risk to management associated with the release of internal reports is increasing, and the intended use of the disclosed information is being questioned more now than before.

---

[34] Except in cases where the report is protected under the attorney–client privilege and/or work product doctrines.

7   AUDIT PROGRAM ISSUES

COMMENTARY ON GETTING A PROGRAM STARTED[35]

How one gets started in the development of an environmental audit program depends to a large extent on who requested the program. Getting started in simplest terms involves the following:

- Gaining management support
- Developing a sound program plan
- Developing a sound launch plan
- Selecting reliable people to perform the audits and training them properly
- Minimizing mistakes

Environmental audit programs typically have their beginnings as program improvement initiatives desired by management, or desired by environmental staff and sold to management. The former situation is typically triggered by an environmental incident at the company or a peer's company, which raises questions in the minds of management as to the quality of their own environmental programs. The later situation, in which environmental staff must sell the program to management, may be triggered by concerns about the quality of the company's programs, but is most frequently associated with continuous improvement efforts. Where management is not already convinced of the need for an environmental audit program, some selling may be required. A convincing presentation will require preparation. Pragmatic managers will want to compare the costs and benefits of the program.

Some benefits include identification of deficiencies so that they can be corrected and management comfort that they are aware of the status of environmental programs. Potential costs may include the following:

- Compensation for in-house auditors
- Fees for an auditing company

[35] This section of Chapter 7 was written by Mr. Ralph Rhodes (see profile preceding Chapter 1).

- Travel costs
- Support costs
- Consultant fees
- Time committed by plant personnel to assisting the auditors
- Cost of correction of deficiencies found
- Penalties awarded by regulatory agencies for reportable deficiencies

A voluntary *Environmental Audit Survey of United States Business,* conducted by Price Waterhouse, L.L.P., to develop benchmark data, targeted a broad range of companies and industries. The range of cost to audit a typical facility was $200 to $150,000 with a mean of $15,401. The annual cost of audit program ranged from $200 to $4,000,000 with a mean of $367,285. The cost at Allied Signal for a typical facility audit is approximately $15,000, and the annual cost of their audit program, which performs 60 audits per year, is approximately $900,000.

Management must understand from the start that there must be a commitment to correct deficiencies found during the audit. Deficiencies found and reported to senior management, but not corrected, pose an unacceptable and more serious personal risk to management. In addition, to focus on specific areas of impact discovered by the audit program, presentations of the audit findings to management should also include critical environmental issues associated with the company's operations and known program weaknesses.

The legal counsel, in its role of protecting the corporation, is a key player in decisions concerning the start-up of an environmental audit program. On the one hand, audit reports pose a risk of self-incrimination. On the other hand, they pose an opportunity to identify program weaknesses so that they can be corrected before they cause significant environmental harm or become a subject for enforcement attention. The legal conclusion as to whether environmental audits are either an opportunity or a risk will depend to a substantial degree on how well informed management is on how environmental audits can be used as an environmental management tool, on how basically conservative the decision making is, and on the effectiveness of the presentation of potential benefits.

Downside risks can be minimized by a close liaison between the audit team and legal counsel. Lawyers may serve on the audit teams, closely monitor all reports, or even serve as a focal point by managing the audits under attorney-client privilege. Just how practical attorney-client privilege really is can be debated, but some programs do employ it.

State environmental audit privilege legislation currently under consideration (or already passed) in many states may raise the comfort level in some. The legislation provides certain protection for audit reports from use in enforcement action. Many informed legal counsels will be inclined to have a positive view of environmental audits on the assurance that deficiencies found will be corrected through a strong, efficient, follow-up system.

Developing a Sound Program Plan — Assuming that management supports the use of auditing, the person or persons responsible for implementing the program must come up with one that will achieve customer (management) expectations, be consistent with the structure and culture of the organization, and avoid downside risks. To achieve all of these objectives, the new environmental program manager may solicit the help of a consulting firm or undertake program development him/herself utilizing some of the excellent information currently available.

An important guide in program design is the recently developed ISO 14012 standard for environmental auditing programs. The standard, a product of the ISO international standards setting process, incorporates principles developed over the 20 years since the environmental audit programs first appeared.

Developing a Sound Launch Plan — Having a good program design, one well suited for the program objectives and the culture of the organization it is expected to serve, is very important. Unless, however, the program is launched with care and foresight, acceptance and impact on the company's environmental, health, and safety (ES&H) program as a whole may not (at least in the early years) match expectations. The importance of management support that is widely communicated cannot be emphasized enough as a key to acceptance of a new audit program. The basic purpose of audit programs, *to evaluate programs and communicate deficiencies to higher management*, makes local management understandably reserved on the subject. Upper management support is needed to assure cooperation at the plant level and to ensure that any deficiencies that are found are promptly corrected. There is an old truism that if you do not plan to correct deficiencies, you should not audit to uncover them. Some companies have reservations about undertaking a program because the large number or severity of deficiencies could be beyond their means to correct. An approach to this situation is to conduct a very broad audit, concluding with an oral report to management. Frequently, the initial survey gives senior management enough confidence in the plant control systems to authorize in-depth audits.

There is another truism to the effect that you should *always do your first audits at a plant where management is supportive.* Even the best prepared, newly formed audit teams will make mistakes that would be very damaging in a hostile plant but only the subject of good-natured banter in a supportive plant.

Early audits will be subject to close scrutiny by the rest of the organization. Errors and oversights are particularly damaging at that point. An experienced consultant or an in-house coach/advisor can be an important damage-prevention/control agent and a source of comfort to the audit program management. The legal counsel will be particularly interested in the early reports. Review and approval of drafts by legal counsel for their opinion, before they are distributed, is an absolute necessity.

In early audits it is particularly important that the audit team carefully explain their purpose and scope to *everyone* with which they interface. The anxiety level is likely to be high at all plant management levels. Misinformation and lack of information are two prime contributors to that anxiety. A clear, complete summary of the audit purpose and scope presented in a friendly, low-key manner will reduce the anxiety and stimulate communication and cooperation.

To the extent possible, higher management should be cautioned not to over react to audit findings. The focus of findings should be placed on corrective action rather than on criticism. This is necessary to build wide support for the audits as a value-added tool to the environmental program rather than a view that the audit program is a basis for management's second-guessing and criticism. Early briefings to site management of this concept will help them understand their role in the program and how they can contribute to its effectiveness.

Selection and Training of People — The preparation for the first audits should include careful selection and thorough training of audit team leaders and members. Audit techniques are not a part of the training of most environmental staff. A one-week course led by experienced audit trainers is a very important preparatory step in all new audit programs. Generally, team leaders require additional training in the specific responsibilities that accompany the leader role. Selection of team members should focus, of course, on technical proficiency, but it should not overlook personality traits that establish good working relationships with plant personnel and thus break barriers to communication that are particularly strong in new audit programs.

Minimizing Mistakes — Success in the early stages requires building a reputation for excellence quickly. Mistakes and oversights, *never acceptable,* are more damaging in the early stages. As mentioned earlier, an experienced coach/advisor is a very desirable participant, particularly in the early audits. Audit plans should be very conservative in the early audits. Allow enough time for each auditor to proceed methodically through his or her assignment with time to focus on both audit procedure and the subject being audited. Frequent review of working papers and discussion of progress/problems among the team members and the team leader(s) are important. Often, it is a good idea to keep the audit scope quite narrow in the early audits, thus minimizing the number of auditors required to do the job and the supervisory workload for the leader(s) and ensuring that everything within the scope is carefully reviewed. Bringing to light issues that others may have missed in earlier, more superficial reviews provides a very important boost for the reputation of the program. Controversial issues should be resolved informally with legal assistance and the involvement of all persons responsible for management of the issue before communication to management. Communication of a "problem" that later is found not to be a problem should be avoided.

Getting started with an audit program is without question a stressful experience to auditor and the audited alike. Attention to the issues discussed in this chapter will reduce stress of all concerned and increase the potential for a smooth launch of a successful audit program. One should *not* follow the advice of the athletic shoe advertisement and "Just do it!"

AUDIT PROGRAM ISSUES

Some fundamental decisions have to be made when initiating or reviewing an audit program. First and foremost are the goals and objectives of the program, because this will set the tone and influence some of the other decisions: whether to develop an in-house program or to use the best available external services; whether the criteria of performance is regulatory compliance or other performance standards, such as corporate standards, best management practice, or beyond-compliance; whether the audit is only to identify issues of action or also to recommend corrective actions; whether there should be a scoring system; and whether the results should be summarized orally or written. All of these factors will be dependent on the goals and objectives of the audit program.

Considerations in establishing an audit program should, as a minimum, address the following issues:

- Goals and objectives
- Schedule, frequency, and types of audits
- Scope
- Choice of development of internal program vs. extensive support types
- Team selection and preaudit preparation
- Facility preparation
- Protection of attorney-client privilege for audit reports
- On-site audit process
- Summary presentation, report style, language, and presentation
- Legal review of audit information and information control
- Fairness and consistency of audit process
- Scoring or ranking system and the use of such systems
- Corrective action plan and readiness plan
- Defined use of audit information for senior management and facility management
- Audit program status report and review

Goals and Objectives — The audit has to relate and meet the overriding goals of the organization. Likewise, there must be upper management commitment. With the exception of a deliberate negligence, the management and employees should be considered blameless of the audit results, as if each audit were a fresh start. However, management accountability should be assigned for corrective action. Maximum benefits of auditing will be attained when the challenges to openness and candor are overcome. The objectives of an audit program should be decided for each company at the onset and may include any of the following:

- To measure effectiveness of environmental management system
- To identify potential liabilities
- To collect data for management decision on resource allocation
- To regulatory and/or corporate performance standard compliance evaluation
- To educate or communicate site and corporate personnel
- To provide data required by outside agencies

The Price Waterhouse survey indicated that the main reasons for auditing are (1) to identify problems internally and correct them before discovery by agency inspections, (2) to improve the company's overall environmental program and make it proactive, (3) assurance that management control systems are functioning, and (4) to decrease the company's operating and financial risks.

Second, factors that would encourage more audits are as follows:

1. Enforcement policy reducing penalties for self-identified, reported, and corrected items
2. Passage of a federal privilege law
3. A presumption against company criminal prosecution if a comprehensive audit program is in place, corrections are made, and violations are reported
4. Enforcement policy eliminates or substantially reduces potential penalties for self-identified, reported, and corrected items reported within a specified period
5. Customers require that a comprehensive audit program be in place
6. Passage of a state privilege law
7. "Fast track" permit procedures are available if a comprehensive audit program is in place (Ref. 1)

An audit is an opportunity for an incoming Plant Manager or Environmental Safety and Health Department Manager to identify and document the existing status of affairs. By identifying all activities and action items, a baseline is established on which to make improvements and against which to measure performance. Such an audit by an incoming manager is instrumental in setting the department's program goals, objectives, and priorities on a schedule. The audit report provides justification for annual budget requests, identifies appropriate staffing needs, identifies target management integration concerns of each departmental function. To incoming managers, a corporate audit to help establish departmental goals and activities is often the most critical step that they request.

AUDIT PROGRAM ISSUES

Characteristics of an Audit — By definition, an audit must be unbiased, periodic, systematic, documented, and conducted by a qualified team. Internal audit programs are under greater pressure for bias due to internal political issues that may lead audit team members to be either antagonistic or biased toward the site host. Audit team selection is also critical in maintaining consistency. When field staff are utilized to support the audit program, it will be important to verify that the field staff have not previously worked at the facility he will be auditing. It has often been noticed that in an audit team made up of company staff and consultants, the consultants have less reservation about critiquing the facility operation and the company staff are more "soft" about mentioning all large and small findings. The frequency of audits should also be fair, depending on the results of previous audits and follow-up of corrective action programs. A defined schedule will help maintain fairness. For example, a repeat finding in two consecutive audits would indicate poor follow-up and should require senior management directives as well as more frequent audits.

A standardized procedure assures systematic assessment such that every facility receives the same scrutiny and any team of auditors using the standardized procedures at the same location would have similar results. An established protocol controls for such bias. Consistency between divisional programs and overseas operations of a large corporation can be maintained through use of protocols, auditor training, and use of feedback evaluation of the auditors and the audit program by the audited facility. Other characteristics of an ideal audit include independence and efficiency (Ref. 20).

The degree and style of documentation is both critical and controversial. While documentation is considered necessary in many companies to ensure management commitment to follow-up, controlled release of information and protection of attorney-client privilege must not be jeopardized.

Professional judgment, interpersonal skills, and experience play a role, and professional certification should be a prerequisite. When auditing is for regulatory compliance, it may seem easy at first, but the interpretive nature and complexity of regulations can lead to long discussions that are sometimes difficult to resolve. In such cases experience and professional judgment play a larger role in conflict resolution. A good interpretation of a regulation with a clear and reasonable approach to compliance can also provide a defense against citations, even though the approach to compliance may be somewhat different from that of the regulators.

Type and Frequency of Audits — How you conduct an audit clearly depends on the type of audit, as well as the objectives of the audit program. There are three major types of audit:

- Regulatory or Corporate Performance Standard Compliance Audit
- Real Property Transaction Audit
- Consent Decree Audit

Audits also fall under three categories — scheduled, "surprise," and reactive audits. In addition to routine scheduled audits and a requested audit by an incoming manager to establish a baseline (a reactive audit), the following should prompt audits:

- Major accidents
- Occupational disease claims above the norm
- Worker's compensation insurance costs increasing above the norm, by >10%
- Regulatory citations
- Previous audit results justify, based on need for immediate action

There are advantages and disadvantages to each of the three categories of audits — scheduled, "surprise," or reactive.

For the scheduled audit, the facility should be told at least 6 months prior, and all requests for information should be initiated at least 60 days prior to the audit. Ample notice allows the facility to prepare with self-evaluation audits and corrective actions. The facility can also schedule infrequent batch productions, arrange for required staff to aide with the audit, and incorporate corrective actions, but short-term fixes and other forms of masking could also be planned.

A surprise audit, by its very nature will have conflicts and resistance. It mirrors a regulatory agency attitude, and required site personnel might not be available. On the other hand, a reactive audit in response to an event is to determine whether the incident was an isolated case or represents the tip of an iceberg. Such an audit should be fully cooperative on both sides and reveal all. Sometimes, a preliminary screening audit can help develop a full audit program. It will help identify the scope.

The frequency of audits range from 1 to 4 years. Some factors used to determine audit frequency for United States sites that are used by E. I. DuPont de Nemours' environmental audit program are presented in Table 7.1 (Ref. 21).

Scope — To define the audit scope, it is necessary to understand facility operation, process description, chemical inventory information, and storage information of raw material, intermediate products, and waste material. Based on the information, the scope of the audit can be identified and may include any of the following topics.

Environment
  Air Pollution and Air Quality
  Spill Control and Emergency Planning
  Polychlorinated Biphenyl Management
  Underground Storage Tanks and Above-Ground Tanks
  Groundwater Protection and Underground Injection Control
  Solid and Hazardous Waste and Shipping
  Water Pollution Control, Industrial and Stormwater Discharges
  SARA and CERCLA Reporting

Table 7.1 Criteria Factors for Audit Frequency

Site characteristics	Every 2–3 years	Every 3 years	Every 4 years
Size and type	Major manufacturing facilities	Small manufacturing, laboratories	Very simple manufacturing, warehouses
Hazardous waste	On-site hazardous substances (permitted TSD facility)	Large quantity generator (>1000 kg/mo)	Small quantity generator (<1000 kg/mo)
Wastewater	Operates an on-site wastewater treatment plant or pretreats before discharge	Discharges processed and sanitary wastes to POTW	Discharges only sanitary wastes
Air	Major sources of air toxins (10 t/yr of 1, or 25 t/yr of a combination). Significant VOC emissions in serious or severe ozone nonattainment areas	Air sources require permits. Significant VOC emissions in moderate or marginal ozone nonattainment areas	No sources require permits
Spill potential	On-site bulk oil or hazardous substances storage of >50,000 gal	On-site bulk oil or hazardous substances storage of 1000 to 50,000 gal	On-site oil or hazardous substances storage of <1000 gal
Groundwater	Documented groundwater contamination and nearby drinking water supply	Suspected prior on-site disposal of potential contaminants — no contamination documented	No prior on-site disposal or documented groundwater contamination

Abbreviations and symbols: < = less than; POTW = publicly owned treatment works; > = greater than; t/yr = tons per year; kg/mo = kilograms per month; VOC = volatile organic compound; gal = gallons.

Pollution Prevention and Waste Minimization Program and Achievements
Permitting, Reporting, Monitoring, and Keeping of Records
Emergency Planning and Community Right-To-Know (EPCRA)
Pesticide Application
Hazardous Waste Operation and Emergency Response

Safety
Process Safety
Loss Prevention and Emergency Response
Employee Safety
Fire Safety
Electrical Safety
Chemical Storage and Handling

Industrial Hygiene
Comprehensive Industrial Hygiene
- Hazard Communication Standard
- Exposure Assessment Program
- Respiratory Protection Program
- Personal Protective Equipment
- Medical Surveillance Program
- Hearing Conservation Program
- Carcinogen/Reproductive Toxin Program
- Employee Training Program
- Access to Employee Exposure and Medical Records

Drinking Water
Radiation Safety
Laboratory Hygiene and Safety
Ergonomics
Biosafety
Asbestos
Lead
Occupational Medicine and Health Surveillance

Other
Product Safety
Chemical Control
Other Toxic Substance Control Act Issues

Team Selection, Team Preparation, Facility Preparation, and Off-Site Preaudit Activities — The preparatory work prior to on-site activities is more extensive than an actual on-site audit. Facility information should include materials and processes, the carcinogens used or stored, the shift schedule, and the names of key personnel to which the audit team will need to talk.

During the preaudit activities the team should have a fair idea of what they will be observing and the kind of issues that are likely to appear. With good preparation, the on-site audit should mostly be a confirmation of information gathered and anticipated issues. Team selection goes hand-in-hand with team preparation because site information is necessary to select the best team.

When selecting a team of auditors, each area of specialty must be covered. Even if an auditor has dual expertise, it is always better to designate one area of his expertise as his responsibility during the audit. For example, an auditor that is both a Certified Safety Professional and a Certified Industrial Hygienist should be given the responsibility for just the safety aspects or the industrial hygiene aspects of the audit because, despite expertise in both areas, there are many details that may be miss during the audit due to the magnitude of both areas being covered. Of course, this would depend on the size of the facility. Team selection must also address internal political issues if auditors are pooled from facility staff. If external auditors are pooled, they must have received an indoctrination or corporate training prior to the actual audit. A preaudit conference is a good time to outline all expectations from all internal and external auditors. In addition to expertise and experience, auditors must have personal attributes to work well with people in what may be confrontational or unwelcome situations. When there is a mix of internal and external auditors, the mix should be consistent because it has been observed that external auditors make more detailed observations than internal auditors and are not afraid to critique the corporate system in place. Finally, although it is standard practice to train new people, it is advisable not to have more than one new auditor under supervision on any audit team.

Team selection is critical to a successful audit. Team leader responsibilities are many and include coordinating the schedules, factual and diplomatic communication with the site manager, managing the team for consistency and fairness of the audit practice, controlling the speed of the audit process to cover the scope intended, conducting formal introductory and closing meetings, and resolving conflicts between auditors and site personnel. With the specialization of auditing, the requirements of auditors and audit leaders are becoming more defined. Most companies require proper formal training and field training. ISO requirements for auditors and audit leaders are specified in Chapter 9.

Facility preparation for an audit should include the following:

- An introductory letter to the site manger to announce the scope of the audit, the items of review during the audit, the audit procedure and reporting style on findings, and a request for designated site personnel for assistance during the audit
- A presurvey questionnaire of issues and concerns as observed by site personnel and site management
- A list of documents for the audit team should include:
  - Corporate and site policies and procedures

- Past audits, site reports, citations, or violations
- Name of site personnel who are responsible for functions and who will be providing information
- Site organization chart with plant manager; process engineer; chemical/equipment purchasing agent; laboratory chemical hygiene officer; environment, safety, and industrial hygiene specialist; medical records manager; etc.
- Facility physical layout map and facility drawings
- Chemical inventory and Material Safety Data Sheet
- Employee training program and documentation
- Safety statistics, accident reports, and environmental reports
- Process description that entails flow chart; material use and production volumes handled; raw materials; intermediate, byproducts, and end products; task description; work shift rotation and seasonal variation; controls for exposure; engineering setup; personal protective equipment used; recent or planned modified process

Like a well-prepared audit team, the best results of audit can be achieved when facility management is cooperative and open, prepared to receive the audit, and ready to act on any findings. Facility commitment to resources should include the following:

- Site personnel to gather data and reports
- Site person to accompany the audit team members during the audit
- Prompt response to findings with corrective actions
- Management attention
- Support for the audit program itself

On-Site Audit Activities — The actual audit at the facility should be as brief as possible to minimize disruption of facility operation; site personnel will then be more cooperative for a shorter duration. Furthermore, when routine audits are approached as an informational and educational session for site personnel, the mood will be conducive to better sharing of information by site personnel. Ideally, an audit should be less than 5 days, preferably 3 days, and the audit should be planned with as much off-site preparation as possible. As stated earlier, the on-site audit should only verify suspicions gathered from preaudit preparation.

The purpose of the introductory meeting is to introduce the audit team; to discuss the purpose, scope, and objectives of the audit; to discuss the schedule and the site staff responsibilities in assisting with the audit; to create an opportunity to establish open lines of communication; and, most importantly, to set the tone of the audit.

A preliminary tour of the facility and its process lines allows the team to identify risk issues not identified or recognized by site personnel. Any findings observed should be noted and verified with site personnel, whether they agree

AUDIT PROGRAM ISSUES

or disagree, and thus there should not be a finding that is a surprise to the facility escort. At the closing meeting all findings should be clarified and identified with a root cause, at the management system level. Any unresolved issues should be noted as such, with arguments being presented by both the audit team and the facility manager, for the legal counsel to review.

Finally, the audit team will need a conference room, clerical and secretarial support, telephone, copier, computer, printer, and software, etc. It is typical for the auditors to bring their laptop and software with readily usable checklists and for the host to arrange for a conference room, telephone, copier, printer, and clerical support.

Unless corrective actions can be triggered by an audit report, the audit program is not ready to embark on actual field audits. A system to follow up and track audit findings should be set up prior to on-site activities; otherwise, you may open yourselves to greater risk of liability.

Audit Report — The language of the report, the distribution of the report, the review by legal counsel for attorney-client privilege, and the liability control of the report are only some of the issues. During the audit program development stage, a decision should have been made on whether the report is to be oral or written, on the written report format, and on who shall receive the final report. A recent voluntary Environmental Audit Survey of United States Business conducted by Price Waterhouse, L.L.P., shows that 96% of the companies interviewed require a formal written audit report, and 68% use a draft of the report to minimize concerns about disclosure. No matter which form of the report is used, 92% of the companies still prepare a written corrective action plan to address findings and file it with the audit report.

Although the closing meeting at the audit site will summarize the findings and the management system issues to all site personnel, the final report is restricted to senior management and to site management personnel responsible for the corrective actions, for purposes of damage control. The report format should consist of preliminary information received prior to the audit, some preliminary issues recognized by the site, management system issues supported by specific findings, and references to site documents. Although the final report is restricted to management, it is important that all site personnel are involved with the audit process and corrective actions thereafter to encourage ownership of environmental responsibilities to each employee.

The language of the report should not contain judgment and alarming adjectives, such as "dangerous" or "lacking." The style of writing should be clear, factual, and precise and should avoid generalizations. The report should unambiguously summarize three important items:

- The compliance status of the facility against federal, state, and local laws and regulations.
- Facility conformance with company corporate policies, procedures, and goals.

- The management system as it exists and its effectiveness in a manner appropriate to assist senior management and site personnel with decision and action plan implementation.

The executive summary, written for senior management, should ideally be one page, and never exceed two. The introduction should contain the who, what, when, where, why, and how of the audit. The management systems findings should be categorized under separate sections for industrial health, safety, and environment, as these organizations may be under more than one umbrella. Each management system finding should be supported by at least three findings. The management system issues are listed in the first column of Table 7.1.

An overall risk ranking supported by reasons should be included in the executive summary. The scoring system should clearly distinguish between findings against regulatory compliance, company policy, or good management practice. Final scoring should be able to rank sites on a relative scale of importance or urgency. This ranking allows prioritization of action items for each facility, as well as sets priorities between facilities.

Finally, there should be positive findings noted as well. The audit report should list the program documents, reports, data, and standard operating procedures reviewed. For interviews conducted, the position titles, *not* the name of the employee, should be noted.

"Attorney-client privilege information" is based on advice from legal counsel. The legal involvement is essential from the development of the program to ensure the attorney-client privilege. Thus all reports should have a legal review and be imprinted with

"ATTORNEY-CLIENT PRIVILEGE CONFIDENTIAL"

This shall set the tone for restricted distribution of reports.

Root Cause Analysis at Management System Level — Finding the root cause of a problem is akin to the identification and treatment of the disease and not just the symptoms. Applying root cause analysis to audits uses the findings (symptoms) to identify the reason behind the problem (disease). To be able to use root cause analysis there must be a thorough, careful, and exacting data collection performed. Root cause analysis has been successfully used in accident and safety analysis and in total quality management.

For identifying root cause at the management system level, a horizontal and vertical analysis of each finding will help in understanding where the majority of the problems lie with each facility. The findings can be charted in a table to summarize the management system level root causes as shown in Table 7.2. The horizontal sections list the audit topics, such as ventilation, ergonomics, hearing conservation, etc. The vertical analysis identifies the management system level issues typically used by auditors, such as Management Support and Management Organization, Line Management Responsibilities,

AUDIT PROGRAM ISSUES

Table 7.2 Summary of Analysis for Root Causes at the Management System Level

Management system level	Audit topic				
	Ventilation	Ergonomics	Hearing conservation	Solid waste	Drinking water
Policy and Procedure				Finding 6	
Management Support and Management Organization					
Line Management Responsibilities		Findings 1, 4, 7			
Actual Operation and Training			Findings 2, 3, 5		
Supporting Documentation and Communication					Finding 10
Keeping of Records Requirements and Access to Records	Finding 23				
Interviews with Employees to Determine Employee Knowledge					

Actual Operation and Training, etc., for each of the horizontal disciplines. This is similar to the ICC and GEMI scoring philosophy under the ESAP, but the ESAP focuses more on the audit of the EMS, while this table focuses more on the tangible findings.

Parallel to GEMI's ESAP, the American Industrial Hygiene Association (AIHA) Occupational Health and Safety Management System (OHSMS) and Guidance Document is the first management system checklist for safety and health, and the document is expected in 1996. The OHSMS is a management system checklist that complements an audit checklist, such as the AIHA's Industrial Hygiene Auditing: A Manual for Practice. There are seven levels of hierarchy used in the example table to determine root cause at the management system level. (The example table is not completely filled in.)

Repeat findings reflect on the effectiveness of the management system. An immediate solution to fix only the findings is only a Band-Aid approach, and similar findings will continue to appear in future audits. The findings only assist in understanding where the management system has failed. The root cause of the problem will be found at the management system level. By targeting a deep-root solution at the management system level, together with the immediate Band-Aid approach, the problem will have a permanent solution.

Audit Program Status Report — How the audit will affect profits, health and safety compliance, public image, and corporate liabilities are the top executives questions. The *ultimate responsibility for confirming closure of findings* rests on the audit program. Although the facility is usually responsible for tracking the closure of findings and reporting the progress to management, the audit program with its subsequent audits is the last step to verify such closure. Therefore the audit program itself should be reviewed for its effectiveness both in identifying root causes at the management system level and at verifying closure of findings during subsequent audits.

There are two types of status reports to senior management: activity reports and evaluation reports of the audit program. The activity report lists the number of audits completed against schedule, audit coverage, and program budgets. Evaluation reports provide an assessment of the overall performance of the organization by including summaries and categorization of audit findings regarding the management system, the reasonably immediate closure of findings, the correlations and differences among different operating units, and the trend analysis. When overall rating is used in the audits, a summary of the overall rating classification is included. The evaluation report also includes the progress made over the year since the inception of the audit program. The reports should be able to answer the questions related to the overall goals of the corporation. The audit program review is usually conducted by feedback from the team leader and members, audited facility staff, and senior management. External auditors often participate to provide a third-party critique of the audit program. They provide the quality control of an internal audit program and are strongly recommended. The program evaluation is also a management tool to ensure quality control measures of the ES&H programs.

This is an essential component of total quality management for continual improvement.

Other Auditing Issues

Multimedia Approach — Audits should focus on the management systems/processes in place, as well as on attaining results (enumerating findings). When an effective management system/process is in place, it should be able to overcome any poor audit results from previous years.

The multimedia approach to auditing has several advantages over a program-specific audit, such as a more comprehensive assessment of the facility with fewer missed violations; the ability to respond more effectively to non–program-specific complaints, issues, or needs; and the development of a better undertaking of cross-media problems and issues. An example of a cross-media issue would be the institution of waste minimization practices throughout a facility. Multimedia pollution prevention audits are typically conducted at larger facilities and require more time on-site and better coordination among audit team members. During a multimedia pollution prevention audit, focus is on the process of looking at all media collectively, rather than specific media releases. During such an audit it is also possible to look at process modification alternatives for pollution prevention strategy.

The environmental programs and the health and safety programs are most often two separate programs under one umbrella. A safety finding is more often an environmental finding as well. At Johnson and Johnson the separate environmental and safety and health auditing programs have about 80% coinciding findings between the two programs. The establishment of two separate programs results in duplication of work, and the audited companies often receive information that is confusing. Often there is a need to create artificial rules, such as spills inside the plant being a safety issue, while spills outside the plant being an environmental issue. The use of an environmental, safety, and health team audit of has many benefits. When there are nonteam audits, if a safety auditor identifies a problem, then an environmental auditor must revisit the area to look at the same issue from an environmental standpoint. By using the ES&H team, the audit requires less labor, the results will be more cohesive for the company, and therefore the combined environmental safety and health audit is strongly recommended. However, this is not an easy task because the current regulations, guidelines, and standards in most instances for the environmental programs and for the safety and health programs have been issued separately.

Scoring and Multiple Performance Standards — Typically, a scoring system is founded on assigning equal weight to each element of the audit scope and on having some type of numerical or categorical ranking for the level of performance for each element. There are two questions regarding this practice: "Should noncompliance with process safety programs weigh the same as noncompliance with environmental programs?" "Should, for example, the process safety program weigh the same in a petrochemical facility as in a furniture assembly plant?" Obviously the risks are not the same in both questions.

This is the same issue that arises when OSHA performs a facility inspection. OSHA is reported to only identify noncompliance checklist items by number, and not the true severity of the noncompliance issue. For auditing, a scoring system exists to rank the potential risk, not just to compile statistics. Therefore, when planning your audit program, look at your scoring system and ensure that it is in fact weighing the risk potential properly and that the scoring system is tailored to each separate facility.

When the scoring results are made available to all plants, comparisons and questions arise. Often company politics cloud the audit issues, diverting attention away from the implementation of the necessary corrective actions. The situation can be further aggravated when the audit team members are from other company sites being audited under the same program. It is recommended that the individual audit reports, particularly the scoring results, be released only to the audited site. However, should the audit discover an exemplary program at one of the audited sites, that program should be recognized for its excellence and shared with the other company facilities. At E. I. DuPont de Nemours and Company there is a classification system, but the purpose is to assist sites in setting priorities for the management of findings, not for comparison of facilities. The focus is then on correction and not on comparative ranking.

Johnson and Johnson's Safety and Health Auditing program does not make the scoring results accessible to all facilities and uses outside consultants exclusively. Any conflicts are resolved internally by the corporate audit program group, who have the knowledge of the overall corporate performance, practice, and history.

Program Revision — In light of the new EPA 1995 policy and the upcoming ISO 14000 EMS final standards, most corporate auditing programs will need to be reviewed to meet the challenges of the nineties. Most programs will need to address the shift from compliance auditing to EMS auditing and the retraining of the auditors. There are many directives and incentives to shift from the traditional compliance auditing to a higher level of auditing at the management system level. The EMS auditing looks for the effectiveness of the program in place and identifies root causes at the program level. The GEMI's ESAP is a good example of an EMS auditing plan. Auditors will require retraining to perform EMS audits. Auditors will need to ask not only the "What?" but also the "Why?" (root cause) of a observation. Knowledge of management concepts will help to achieve the shift from compliance to management system auditing. In turn, site personnel and plant managers will need to be informed of this shift of focus in the auditing program and process.

Management and Employee Teamwork — There is a continuing need for greater management commitment. In addition to policy statements, senior management's direct involvement with the auditing program may make a significant difference in how the audit program is viewed by company personnel. Management needs to foster employee involvement and public awareness. Secrecy or lack of information causes undue anxiety. Sharing of information

AUDIT PROGRAM ISSUES 101

about the company's environmental performance, goals, and priorities with the employees will lead to greater ownership of their responsibilities. The public has a right to know about the safety of their environment, and the environmentally conscious society keeps up with the good deeds and bad deeds of the players. Resourceful companies and organizations have already effectively utilized public communication to their advantage through green reports and green marketing to increase consumer and public approval.

Process-Oriented Focus — The traditional audit programs tend to measure levels of performance only. An EMS audit will focus on the "systems or processes" in place to determine the progress made to fix findings, even though the desired performance may not be a reality. A process-oriented focus supports the fundamental concepts of EMS auditing, and its use can better predict overall environmental performance.

Third-Party Review — An independent third-party review of the EMS is suggested by the ISO 14000, the ELP's pilot projects, and other initiatives. A third-party review will provide a reality check in many ways. The outside review can give a company an objective look at its EMS, identifying weak or nonexistence program elements that were missed by internal staff due to overfamiliarity. Third parties are also not politically motivated and will provide an unbiased and unrestricted evaluation. Experienced third-party auditors will be able to provide insight gleaned from the other programs and sound advice on EMS improvement. Third-party review is a quality control measure, especially necessary for companies or organizations utilizing internal staff exclusively. Some of the leading companies have already conducted a third-party review to identify program needs to prepare for ISO 14000 registration.

8 AUDIT-RELEVANT REGULATIONS

Before an audit can be properly planned, all state and federal laws and regulations that apply to the facility must be understood.[36] In the past many ES&H laws and regulations applied only to industry and others only to government. Today through the issuance of executive orders and new legislation, the ES&H laws and regulations apply to the federal government. However, there are still laws that apply only to the government, such as the National Environmental Policy Act (NEPA). NEPA requires the government to prepare a detailed statement for every major federal action that affects the environment. Industry is not required to comply with NEPA unless it is participating in a project that has federal involvement. Although there is no NEPA parallel at the federal level, there are laws in some states, e.g., New York and California, that require environmental documentation for major projects. It is this type of example that necessitates auditors to be aware of the laws and regulations and how they pertain to their facility.

The following is a brief overview of some of the federal laws that may effect an audit. The list is not all inclusive, and the pertinent state regulations are too numerous to list. It is the auditors responsibility to be aware of the federal statutes and to check with their respective state for relevant state statutes and issues. Always discuss the applicability of the statues with legal counsel before planning your audit.

OCCUPATIONAL SAFETY AND HEALTH ACT

The Occupational Safety and Health Act of 1970 established the framework for the development of nationwide workplace safety and health standards. The act authorized the establishment of the Occupational Safety and Health Administration (OSHA). OSHA is responsible for establishing the standards that ensure that the workplace is a safe place. The requirements may be in the

---

[36] This chapter was written by Mr. John Falkenbury (see profile preceding Chapter 1).

form of prescribed training, practices, methods, standard operating procedures, or standards for the company.

Some of the OSHA's regulations specifically prescribe compliance audits, while others only imply the use of audits. For example, OSHA's Process Safety Management standard requires compliance audits to be performed at least every 3 years to prevent accidental releases.

CLEAN AIR ACT

The Clean Air Act (CAA) evolved from the Air Quality Act of 1967. Major revisions to the original act were passed in 1970, 1977, and 1990. The last revision is known as the Clean Air Act Amendments (CAAA). Under this act new and existing point sources, ambient air quality for attainment and nonattainment areas, visibility impairment, hazardous air pollutants, acid deposition are all regulated.

The states are required to maintain the ambient air quality within the state at levels consistent with the national ambient air quality standards (NAAQS). The EPA has established six NAAQS: sulfur dioxide, particulate matter (10 microns or less), ozone, lead, nitrogen dioxide, and carbon monoxide. The states have to issue EPA-mandated State Implementation Plans (SIP) for those areas designated by the EPA as in nonattainment with the NAAQS. The SIPs limit the emissions from specific sources in areas of non-attainment to meet the NAAQS. The strictness of the SIP depends on the attainment within the area.

When the CAAA was enacted in 1990, it contained a provision requiring the EPA to promulgate rules to ensure that actions taken by federal agencies "conform" to the appropriate SIP. The conformity rule is to ensure that a federal action does not contribute to any violations of the NAAQS.

CLEAN WATER ACT

The Clean Water Act (CWA), formerly known as the Federal Water Pollution Control Act (1972), was enacted to " . . . restore and maintain the chemical, physical, and biological integrity of the Nation's waters." The CWA has five major segments:

- The setting of national minimum effluent standards
- Water quality standards
- A discharge permit program
- Provisions for special problems, e.g., oil spills
- A construction loan program

The discharge permit program is known as the National Pollutant Discharge Elimination System (NPDES). Although NPDES is a federal program with federal discharge limitations, the states have jurisdiction under the auspices of the EPA.

The latest proposed CWA amendments are reported to tie the discharge permits issuance to audit requirements.

OIL POLLUTION CONTROL ACT

In 1990 the Oil Pollution Control Act of 1990 revised the oil and hazardous substance discharge requirement of the CWA by making prevention, removal, and restoration high priorities with strong enforcement provisions.

COMPREHENSIVE ENVIRONMENTAL RESPONSE, COMPENSATION, AND LIABILITY ACT

The Comprehensive Environmental Response, Compensation, and Liability Act (CERCLA) was enacted in 1980 in response to a growing national concern about the release of hazardous substances. CERCLA provides for liability, compensation, cleanup, and emergency response for hazardous substances released to the environment and for the cleanup of inactive hazardous waste disposal sites. The act requires the identification, characterization, and cleanup of inactive hazardous waste sites by responsible parties and imposes response and reporting requirements for operations from which hazardous substances have been released. The Superfund Amendments Reauthorization Act (SARA) amended many provisions of CERCLA. SARA increased the revenues for cleanup and clarified many areas that were unclear in the original legislation. However, CERCLA still focuses on the cleanup of inactive hazardous waste sites and the distribution of the cost of the cleanup among the principal responsible parties.

EMERGENCY PLANNING AND COMMUNITY RIGHT-TO-KNOW ACT

The Emergency Planning and Community Right-to-Know Act (EPCRA) was part of SARA. EPCRA requires states and local communities to prepare plans for dealing with emergencies relating to hazardous substances. The three major segments of EPCRA are emergency planning and notification, reporting requirements, and general provisions. Under the second segment, reporting requirements, there are three distinct provisions for reporting on two different groups of chemical substances. The first two provisions require inventory reports on hazardous chemicals, i.e., those for which there is a Material Safety Data Sheet. The third provision requires annual reporting to the EPA and your state on environmental releases of listed toxic chemicals (in excess of threshold quantities) manufactured, processed, or otherwise used at your facility.

POLLUTION PREVENTION ACT

The Pollution Prevention Act (PPA) of 1990 establishes a national policy for waste management and pollution prevention to improve environmental quality. The policy focuses on source reduction for waste prevention, followed by recycling, treatment, and, as a last resort, disposal or release. Incorporation of pollution prevention activities offers substantial savings in reduced raw materials, pollution control, and liability costs, as well as protects the environment and reduces the risks to worker health and safety.

The PPA contains provisions for the development and implementation of a source reduction strategy. Such provisions include the establishment of standard methods of measuring source reduction and an advisory panel for data dissemination efforts, the identification of source reduction goals and barriers, and the development of training and awards programs. Additional stipulations are made for awarding grants to state programs that promote the use of source reduction in businesses and for a source reduction clearinghouse to serve as a pollution prevention technology transfer center.

The PPA requires facilities that file an annual toxic release form also to file a toxic chemical source reduction and recycling report for the preceding year. The EPA biennial report that is submitted to Congress includes an analysis of trends in source reduction in industry; identification of barriers to achieving source reduction; identification of industries and pollutants that require priority assistance in source reduction; recommendations for incentive programs, research, and development; and evaluation of data gaps and duplications.

RESOURCE CONSERVATION AND RECOVERY ACT

The Resource Conservation and Recovery Act (RCRA) governs the safe treatment and disposal of waste and regulates waste management practices for generators, transporters, and owners and operators of waste TSD facilities. RCRA, an amendment to the 1965 Solid Waste Disposal Act (SWDA) in 1970, was designed to regulate hazardous and solid wastes from cradle to grave. RCRA was amended in 1976 and again in 1984. The 1984 amendment is known as the Hazardous and Solid Waste Amendments (HSWA). The HSWA profoundly changed the 1976 act by expanding its scope and increasing its level of detail in many of its provisions. As part of its expanded scope, additional provisions were added: the regulation of underground storage tanks (UST), medical waste tracking programs, state and regional solid waste management plans, etc.

TOXIC SUBSTANCES CONTROL ACT

The Toxic Substances Control Act (TSCA) of 1976 provides requirements to safely regulate the manufacture, processing, distribution in commerce, use, or disposal of chemical substances and mixtures that may present an unreasonable risk to either the public health or the environment. The act, amended three times, gives regulatory authority to the EPA. TSCA places the responsibility on the manufacturer to provide the EPA with the ES&H data on the chemical substances and mixtures they produce. One group of the well-known chemicals regulated by TSCA are polychlorinated biphenyls.

ENDANGERED SPECIES ACT

The Endangered Species Act (Title 16 U.S.C. 1531, et seq.) of 1973 established a program for the conservation of endangered species and their ecosystems. Together with the Fish and Wildlife Coordination Act it provides for the designation and protection of plant and animal species that are threatened or endangered. The Fish and Wildlife Coordination Act authorizes the Secretary of the Interior to provide assistance to and cooperate with public and private organizations in the development and protection of the Nation's fish and wildlife. The Department of the Interior publishes the Fish and Wildlife Service's List of Endangered and Threatened Wildlife and Plants in Title 50 CFR Part 17. Although your facility may not have an endangered species on its premises, its environmental releases may be affecting an endangered species. Further, your property may be within an area that is designated as an endangered species habitat.

OTHER ACTS

The following are very brief descriptions of additional acts that may affect an audit.

Safe Drinking Water Act (SDWA) — The SDWA authorizes the EPA to promulgate regulations under two specific programs: the first protects the Nation's public drinking water supplies; the second protects subsurface waters. The SDWA regulates contamination threats to groundwater and has established a national program for the protection of drinking water sources by the development of primary drinking water standards known as Maximum Contaminant Levels.

The Federal Insecticide, Fungicide, and Rodenticide Act — This act, as amended, authorizes the EPA to promulgate regulations governing the use and disposal of pesticides.

National Historic Preservation Act — This act establishes the policy of the United States government to protect and preserve historic structures, sites, and artifacts.

American Indian Religious Freedom Act — This act establishes a policy of the United States government to protect and preserve for American Indians their inherent right of freedom of religion, including access to sites.

Noise Control Act — This act establishes a means for coordinating federal noise control research, setting noise emission standards, and providing information to the general public.

Coastal Zone Management Act — This act establishes and supports national coastal zone management policies.

Marine Protection, Research, and Sanctuaries Act — This act regulates the dumping of materials into ocean waters.

Wild and Scenic Rivers Act — This act, as amended, establishes a national wild and scenic rivers system to preserve and protect selected rivers of the nation.

Hazardous Materials Transportation Act — Under this act the Department of Transportation regulates the packaging and transportation of hazardous material.

NATIONAL ENVIRONMENTAL POLICY ACT AND ENVIRONMENTAL QUALITY IMPROVEMENT ACT

The National Environmental Policy Act (NEPA) was one of the first environmental protection statutes, and it focuses on activities of the federal government. The act established the Council on Environmental Quality (CEQ) to advise the president on environmental policy matters.

Under NEPA, federal agencies and others using federal funds must conduct an Environmental Assessment (EA) to determine whether the action in question will have a significant impact on the environment. If the EA concludes that there will be a significant impact, then an environmental study must be performed and an Environmental Impact Statement (EIS) written. And if the EA concludes that there will not be a significant impact, then the agency must issue a Finding of No Significant Impact.

The EIS must evaluate the impacts of the federal action and its alternatives on the environment. All reasonable alternatives available to the agency or agencies involved, including a No Action Alternative, must be evaluated in the EIS. The EIS is a tool for making decisions. The agency must use the EIS in its decision on whether to proceed with the proposed action or with one of its alternatives, including a No Action Alternative. Once the agency has decided its course of action, it must prepare a Record of Decision addressing all significant alternatives and issues raised during the NEPA process.

The Environmental Quality Improvement Act places the primary responsibility for implementing policies for the enhancement of environmental quality with the state and local governments. It also established the Office of Environmental Quality to provide administrative support for the CEQ so that it would have the resources to carry out its responsibilities under NEPA.

Council on Environmental Quality Memoranda — The CEQ has, in the past, issued memoranda in the *Federal Register* that are pertinent to NEPA and the development of NEPA documents. The memoranda are on the subjects of wild and scenic rivers and prime agricultural land. The memoranda follow.

Interagency Consultation to Avoid or Mitigate Adverse Effects on Rivers in the Nationwide Inventory — The memo is intended to assist agencies in meeting the responsibilities under the president's directive, National Wild and

Scenic River System. Each agency is responsible, as part of its normal planning and environmental analysis under NEPA, for avoiding or mitigating adverse effects on rivers identified in the nationwide inventory prepared by the Heritage Conservation and Recreation Service (HCRS). The development of rivers out paces the ability to protect rivers that qualify for inclusion in the National Wild and Scenic River System. The rivers identified in the nationwide inventory could be depleted before they are fully assessed for qualification. Therefore the federal agency must limit the adverse affects on these rivers.

The directive requires that the agencies study all alternatives before acting on those identified rivers. Where agency actions could effectively foreclose the designation of a wild, scenic, or recreational river segment, the president has directed the agency to consult with HCRS. Consultations with HCRS in early planning stages may reduce resource management conflicts and encourage early resolution of problems.

Analysis of Impacts on Prime or Unique Agricultural Lands in Implementing the National Environmental Policy Act — Approximately 1 million acres of prime or unique agricultural lands are being converted irreversibly to non-agricultural uses each year. On August 30, 1976, the CEQ issued a memo recommending analysis of prime or unique farmlands in preparation and review of environmental impact statements (EIS). Since issuance of the memorandum, EISs indicate that Federally funded projects may be directly or indirectly adversely affecting farmland. These impacts have not been accounted for during environmental assessments (EA). Federal agencies can substantially improve the efficiency of impact analysis by closely following the National Environmental Policy Act (NEPA) regulations (40 CFR 1500–1506, November 29, 1978). The effects of an impact must be an integral part of the EA process and a factor in the decision on whether to prepare an EIS. Alternatives should be considered to avoid or mitigate adverse effects during the scoping process. The United States Department of Agriculture will cooperate with all agencies needing any technical data or assistance in planning projects or developments, in assessing impacts on prime or unique agricultural lands, and in defining alternatives.

The CEQ urges all agencies to apply the goals and policies of NEPA in planning their actions so that the limited resources from agricultural lands can be maintained.

Prime and Unique Agricultural Lands and the National Environmental Policy Act — To review agency progress or problems in implementing NEPA, the CEQ requests periodic reports from federal agencies. The memo includes details to be included in the first report (due by November 1, 1980), such as the following:

- Summary of agency policies, regulations, and other directives specifically for protection of prime or unique agricultural lands.
- Specific impact statements and other documents covering activities having a direct or indirect impact on these lands.

EXECUTIVE ORDERS

Executive orders are issued by the president and affect the way the federal government will conduct business. They will have no bearing on the private sector, unless the company is involved with a government contract or government-owned property. Contact the federal agency administrating your project to ascertain if or how these orders may effect your audit.

Executive Order 11988, *Floodplain Management,* was issued to avoid to the extent possible the long- and short-term adverse impacts associated with the occupancy and modification of floodplains development wherever there is a practicable alternative.

Executive Order 11990, *Protection of Wetlands,* was issued to avoid to the extent possible the long- and short-term adverse impacts associated with the destruction or modification of wetlands and to avoid direct or indirect support of new construction in wetlands wherever there is a practicable alternative.

Executive Order 12088, *Federal Compliance with Pollution Control Standards,* ordered each agency must comply with applicable pollution control standards, such as TSCA, CWA, SDWA, CAA, Noise Control Act, SWDA, Atomic Energy Act/Radiation Protection Guidance, Marine Protection Research and Sanctuaries Act, and Federal Insecticide, Fungicide, Rodenticide Act. The agencies were also instructed to cooperate with the EPA, State, interstate, and local agencies in prevention, control, and abatement of environmental pollution.

Executive Order 12856, *Federal Compliance with Right-To-Know Laws and Pollution Prevention Requirements,* directed federal facilities to set an example for the rest the country and become the leader in applying pollution prevention to daily operations, purchasing decisions, and policies. All federal facilities must comply with the Emergency Planning and Community Right-to-Know Act (EPCRA) and the Pollution Prevention Act of 1990.

Executive Order 12898, *Federal Actions to Address Environmental Justice in Minority Populations and Low-Income Populations,* was issued to ensure that federal activities do not place low-income and minority communities at disproportionate risk by requiring that agencies identify the potential economic and social impact of actions in accordance with health and environmental laws. The executive order also addresses the issues of public involvement and data collection efforts to ensure that such efforts consider the impacts of environmental hazards on low-income and minority communities.

9 INTERNATIONAL STANDARDS ORGANIZATION 14000 REQUIREMENTS FOR AUDITING

The International Standards Organization (ISO) is an international voluntary consensus organization based in Geneva, Switzerland, founded in 1946 to develop international standards and reduce trade barriers resulting from national standards. The ISO is most well known in the management system context for the ISO 9000 Quality Management Standards, adopted in 1987 by the United States, Canada, the European Community, and other countries. The companies with ISO 9000 certification/registration carry a stamp of approval for their corporate quality management system that meets the ISO standards. This certification provides a competitive edge to the company by assuring that a system has been implemented that is designed to achieve quality products and services. Many customers look for this type of assurance when selecting a vendor or purchasing a product. Similarly, ISO 14000 registration is expected to provide a competitive edge to registered companies by providing the assurance of a proactive internal EMS and by creating an improved public image.

The ISO 14000 is a proposed set of standards for environmental management that focus on management evaluation and product evaluation. A Technical Committee (TC) 207 for the ISO 14000 was set up to develop a series of international standards relating to the EMS. The TC 207 has six subcommittees and one working group under its direction. The subcommittees and work group are as follows:

Environmental Management System Specification
Environmental Auditing
Environmental Labeling
Environmental Performance Evaluation
Life Cycle Assessment
Terms and Definitions
Environmental Aspects in Product Standards Work Group

The subcommittees on Environmental Management System, Environmental Performance Evaluation, and Environmental Auditing promote management evaluation and the subcommittees on Life Cycle Assessment, Environmental Labeling, and Environmental Aspects in Product Standards promote product evaluation.

Many globally competitive corporations are preparing to register for the ISO 14000 and to use the registration for green marketing. The ISO 14000 registration requirements contain criteria for environmental management system auditing. ISO 14000 is a voluntary program and if a corporation chooses to register, ISO 14001 outlines the requirements for registration and the other 14000 series are guidance documents.

In addition to the ISO 14000 EMS, there are other established voluntary international EMS standards. The United Kingdom has been a leader with the development of the first international standard, British Standard 7750. The Eco-Management and Audit Scheme is a European EMS program based largely on the British Standard 7750. These are voluntary programs that overlap the ISO 14000 program elements. Another recognized EMS standard is the ICC Business Charter EMS. A comparison of the various EMS standards may be found in Table 9.1[37] (Ref. 22).

The EPA has submitted comments on the ISO 14000 standards. Although not all the comments were incorporated, the EPA is generally satisfied with the pollution prevention and compliance auditing language that was included. It is essential to obtain EPA's sanction of ISO 14000 to gain credibility among future registrants, because an EPA sanction would automatically lead to confidence in ISO 14000 registration by United States companies, and industry would prefer not to have to answer to multiple standards.

The American National Standards Institute is the United States representative to the ISO for the development of the final EMS standards. The ISO 14001 EMS specification is currently a Draft International Standard (DIS) that, along with four supporting guidelines, is eligible for adoption as a national standard. The United States procedure for adopting the Draft International Standard as a national standard is called "synchronization." The synchronization by the institute is expected to be completed by early 1996, while the adoption of the final international standards are expected in mid-1996.

The draft EMS standard is not prescriptive in that it does not impose end-of-pipe emission control, but rather evaluates the *quality of the process*. At this time ISO 14000 is a DIS and only covers environmental issues, and discussions are under way for the development of ISO standards on occupational health and safety and for a combined way to audit EMSs and quality. The ISO Strategic Advisory Group of Environment Vision is requesting an integrated management system with standards on quality, finance, environmental and occupational health and safety, and others.

---

[37] Antoinette Bang of ABB Environmental Services.

Table 9.1 Comparison of Environmental Management Systems

ICC Business Charter	Eco-Management and Auditing Scheme	British System 7750	ISO 14000 (TC 207)
1. Corporate Priority	1. Policy, objectives programs; prevent or eliminate pollution, cooperate with public, information to public	4.2 Environmental policy: relevant, understood, publicly available	4.1 Environmental policy: relevant commitment to continual improvement and pollution prevention
2. Integrated Management	2. Organization and personnel: responsibilities and authority, management representative	4.3 Organization: all functions that have, or could have, a significant effect	4.3.1 Structure and responsibility: roles, responsibility, and accountability; documented management communication
3. Process of Improvement	6. Environmental audits 1. Environmental objectives and programs	4.5 Environmental objectives and targets 4.11 Management review	4.2.3 Environmental objectives and target 4.4.1 Monitoring and measurement 4.5 Management review
4. Employee Education	2. Organization and personnel: communications and training	4.3 Organization: communications and training	4.3.2 Training, awareness, and competence: all levels: roles responsibilities, potential consequences
5. Prior Assessment	4. Operational control; approval of planned processes and equipment	4.6 Environmental management program: related to new installations or modifications of processes	4.2.4 Environmental management program still to be discussed with ISO 207/Subcommittee I

Table 9.1 Comparison of Environmental Management Systems (Continued)

ICC Business Charter	Eco-Management and Auditing Scheme	British System 7750	ISO 14000 (TC 207)
6. Products and Services		4.6 Environmental management program: development of new products or services	4.2.1 Environmental aspects: of its activities, products, and services
7. Customer Advice			4.3.3 Communications: significant environmental
8. Facilities and Operations	3. Environmental effects and registration 4. Operational control	4.8 Operational control: procedures, monitoring, documented work procedures	4.3.5–6 Document and operational control: documented control procedures routing operations
9. Research and Development		4.6 Environmental management program: development of new products or services	4.2.4 Environmental management program
10. Precautionary Approach			4.4.2 Nonconformance and corrective and preventative actions

ISO 14000 REQUIREMENTS FOR AUDITING

11. Contractors and Suppliers	3. Environmental effects and registration: procedures dealing with procurement	4.8 Operational control: procedures, monitoring, documented work procedures	4.4 Checking and corrective actions: Environmental aspects of goods and services used by the organization
12. Emergency Preparedness	3. Environmental effects and registration: incidents, accidents, and potential emergency situations	4.4 Environmental effects 4.8 Operational control 4.9 Environmental records	4.3.7 Emergency preparedness 4.4.2 Operational control 4.4.2 Nonconformance action
13. Transfer of Technology			
14. Common Effort			
15. Openness to Concerns	5. Environmental documentation and records 7 Environmental statement	4.2 Environmental policy 4.4 Environmental effects register	4.1 Environmental policy 4.3.3 Communication
16. Compliance and Reporting	6. Environmental audit and review	4.10 Environmental audits 4.4 Environmental effects register	4.4.4 EMS-Audit 4.3.3 Communication
	5. Environmental documentation and records: describes the interactions of system elements	4.7 Environmental management manual	4.3.4 Environmental 4.3.5 Documentation

The ISO 14000 standards for auditing are divided into three groups: auditing principles, auditing procedures, and auditor qualifications. At this time the TC 207 subcommittees have developed the following guidance documents relevant to EMS:

- ISO/DIS 14001 EMS Specification Standards — Specification with Guidance for Use
- ISO/DIS 14004 General Guidelines on Principles, Systems, and Supporting Techniques (Guidance and Interpretation of 14001)
- ISO/DIS 14010 Guidelines for Environmental Auditing — General Principles for Environmental Auditing
- ISO/DIS 14011 Guidelines for Environmental Auditing — Auditing Procedures
- ISO/DIS 14012 Guidelines for environmental auditing — Qualification Criteria for Environmental Auditors
- ISO/ 14013 Management of Environmental Audit Programs

Each nation adopting ISO 14000 will need to develop its own scheme for the accreditation of registrar companies, for the certification of individual auditors performing third-party audits of EMSs, and for the approval of training programs offered to auditors seeking certification.

A proactive EMS, consistent with sustainable development, must be integrated in operations, finance, quality, and marketing. Potential benefits associated with an effective EMS include the following:

- Meeting customer environmental expectations
- Maintaining good public and community relations
- Satisfying investor criteria and improved access to capital
- Obtaining insurance at reasonable cost
- Enhanced image and market share
- Meeting vendor certification criteria
- Ability to dispose of waste
- Cost control
- Liability limitation
- Demonstration of due diligence
- Conservation of input materials and energy
- Easier site selection and permitting
- Technology development and transfer
- Improved industry-government relations
- Improved environmental performance and state of the environment
- Agency recognition
- Passport to emerging market
- Contractor performance criteria

ISO 14000 REQUIREMENTS FOR AUDITING

A corporation with a proactive EMS has a competitive edge due to operational and production efficiencies, loss elimination, and increased acceptability of their products and services in the market. To implement an efficient EMS, an organization should have the support mechanisms necessary: the technical, nontechnical (human), and financial resources. An effective EMS should be integrated with existing management elements including the following:

- Resource allocation
- Operational controls and documentation
- Information and support systems
- Training development
- Organization and accountability structure
- Reward and appraisal systems
- Measuring and monitoring systems
- Communication and reporting

Discussions are currently underway to evaluate harmonizing the ISO 9000 and ISO 14000 series to allow for combined EMS and quality management system audits and for the development of ISO standards on occupational health and safety. In the United States the American National Standards Institute (ANSI) is sponsoring an open meeting in 1996 to discuss the issue of "International Standardization of Occupational Health and Safety Management Systems — Is there a Need?" The American Industrial Hygiene Association (AIHA) has requested to be the secretariat for the Occupational Health and Safety Management System guidance development. The AIHA already has in progress a third draft of a document titled *Occupational Health and Safety Management System and Guidance Document.* This document is a management system checklist for safety and health audits.

REQUIREMENTS OF AN AUDIT PROGRAM

The ISO develops two basic types of generic standards, guidance documents, and specification documents. All of the 14000 series documents are guidance documents except for ISO 14001, which is a specification (*Environmental Management Systems — Specification With Guidance for Use*). The draft standards on auditing (ISO 14010.2, ISO 14011/1.2, and ISO 14012.2) may be found in Appendix B.

Guidelines for Environmental Auditing — For the execution of audits, the guidance document ISO/DIS 14010 *Guidelines for Environmental Auditing — General Principles,* provides the following:

- Definitions of audits, audit criteria, audit findings, audit teams, and related terms
- Requirements for an environmental audit

- General principles, including the standards of due professional care, objectivity, independence, and competence
- Discussion of systematic procedures, the development of audit criteria and evidence, and the reliability of audit findings and any conclusions
- A list of audit-related information that may be included in reports
- Responsibilities of the audit team and the relationships between auditors, the audited, and clients.

Procedures for Auditing — The guidance document ISO/CD 14011 *Guidelines for Environmental Auditing,* establishes audit procedures that provide for the planning and performance of an audit of an EMS. The procedures include the following;

- Specific terms and definitions not found in 14010
- Description of audit objectives
- Clarification of roles and responsibilities within the audit team
- Description of the audit process[38]
- Discussion of audit report preparation, content, and distribution.

AUDITOR QUALIFICATIONS

In ISO/DIS 14012 *Guidelines for Environmental Auditing — Qualification Criteria for Environmental Auditors,* the requirements for education, training, experience, and professional personal traits for both an auditor and a lead auditor are specified. The requirements apply to both internal and external auditors. The ISO standards for auditors are for two levels, auditors and lead auditors, and their qualification criteria are described below. However, please note that at this time these general qualifications may not be all that will be required to become a certified auditor.

General Qualifications — The general ISO qualifications for an auditor are as follows:

- Appropriate work experience
- Education, a relevant degree with two years experience, or four years of work experience without a relevant degree
- Auditor training, both formal and on-the-job training (four audits and 20 workdays within a period of not more than 3 years) under the supervision of a lead auditor to acquire competence with audits
- Personal attributes and skills in communication, interpersonal skills conducive to diplomatic teamwork and tact, independence and

---

[38] The description includes the scope, document review, preaudit preparation, audit planning and team assignments, working documents, executing the audit, collection of evidence, development of audit findings, and initial and closing meetings.

objectivity to accomplish auditor responsibilities, and personal organizational skills for the effective and efficient performance of the audit.

The formal training received by an auditor should include the following:

- Environmental science and technology
- Technical and environmental aspects of facility operations
- Relevant requirements of environmental laws, regulations, and related documents
- Environmental management systems and standards against which audits may be performed
- Audit procedures, processes, and techniques

Lead Auditor Qualification — Additional ISO requirements for the lead auditor are as follows:

- Fifteen workdays and three audits within 3 years and demonstration of personal attributes and skills for leadership in the audit process
- Maintenance of competence, due professional care, code of ethics, and respect for confidentiality
- Language requirements — support of a person with language skills who is not subject to pressure

TRAINING

The training of auditors has always been an important component of the auditing program and will need revisiting with the advent of ISO 14012 *Auditor Qualification Guidelines.* The ISO 14012 sets the requirements for EMS auditing. At this time there is no single accepted certification or registration program for environmental auditors in the United States.

The Environmental Auditors Registration Association of the United Kingdom has an auditor certification and training course accreditation program and is performing for groups in other countries. For ISO 9000, the Registrar Accreditation Board in the United States, in conjunction with ANSI, accredits the registrars that register companies under ISO 9000, accredits training programs that train the auditors, and certifies the auditors. It is important to note that the trainers do not certify auditors as many people might assume, but the Registrar Accreditation Board does. As the ISO 14000 is being modeled after ISO 9000, the registration, accreditation, and certification for ISO 14000 will probably take a similar approach (Ref. 23).

In the United States there will be two systems, ANSI's National Accreditation Program and the Registrar Accreditation Board's National Accreditation Program. The main difference is the Registrar Accreditation Board will still directly certify auditors, while the ANSI will accredit certifying bodies. Both will accredit registrars and approve training programs.

Because of the shift from the conventional compliance audit to the audit of the management systems, training will require a similar shift. Many auditors who come from a technical background will need to go beyond the "what" of a problem, to identify the "why" instead. Many auditors have found difficulty with this shift. Therefore many aspects of training are expected to change with the introduction of ISO 14012 guidelines on auditor qualifications.

The ISO 14012 does not mention any requirement for a mock audit as part of the training program. The Association's accredited courses require at least one practical exercise/site visit. Although this is considered a difficult element for most training organizations, the more difficult one will be the evaluation of the "personal attributes" required under ISO 14012.

The "personal attributes" criteria under the ISO 14012 include, but are not limited to, the following:

- Competence in clearly and fluently expressing concepts and ideas, orally and in writing
- Interpersonal skills conducive to the effective and efficient performance of the audit, such as diplomacy, tact, and the ability to listen
- Ability to maintain independence and objectivity sufficient to permit the accomplishment of auditor responsibilities
- Skills of personal organization necessary to the effective and efficient performance of the audit
- Ability to reach sound judgments based on objective evidence.

Some auditors may have difficulty dealing in all types of situations. However, they can be valued team members, on certain types of audits, when teamed with the right partner. It may be a question of a successful team selection, not a matter of the lacking of a personal attribute. In other situations some auditors may do well when provided some guidance and coaching in conflict resolution. In the future, documentation of the number of audits conducted and documentation of adequate training will be important for certification of auditors.

10 SECURITIES AND EXCHANGE COMMISSION

ENVIRONMENTAL ADVANTAGES

The choice between minimum compliance and maximum disclosure is a fundamental theme to consider and is shaped in large part by the culture of your company. Ideally, companies should address environmental liabilities while informing the public of the steps it is taking to prevent, mitigate, and eliminate environmental problems. However, realistic concerns do not allow many organizations to practice in such a forthright manner. An annual report can be a powerful means of a company's overall communication to investors and the general public. This means of communication can be further enhanced by the publication of a green (environmental) report. Failure to mention any environmental issues would not sit well with the environmentally educated public that exists today. In fact, it would be propitiously proactive to use the annual report supplemented by a green report, as an opportunity to communicate information about the company's environmental programs to investors, customers, and employees and to achieve a more positive pubic image.

Many companies have seized the opportunity to use environmental issues to enhance their public image by such avenues as environmental advertising and green marketing. For example, ISO 14000 registration is considered a excellent green marketing tool in the global market and distinguishes the registered company from competitors. Some prominent examples of integrating environmental and business programs are listed below (Ref. 24):

- Patagonia's long-term goal was to reduce its impact on the environment at every level of its manufacturing and distribution processes. In addition, the company has donated 10% of the pretax profits to environmental groups.
- Nike recently announced that they will begin a pilot program to recycle sneakers.

- The 3M Company sold scraps from the production of various films to companies that manufacture clothes hangers, spatulas, and other plastics items, resulting in the prevention of nearly one billion pounds of releases to air, water, and land. This saved 3M about $480 million in waste material disposal and pollution control costs.
- Eastman Kodak Company reengineered its single-use camera so it could be sent back to Kodak for recycling. Since its inception in 1988 more than three million cameras have been returned to Kodak for recycling, diverting more than 500,000 pounds of material from its waste stream.

INDUSTRY PRACTICE AND REQUIRED DISCLOSURES

There is a growing interest by the SEC, corporate management, and shareholders in integrating the environmental performance and liability information into the annual financial report. The number of Standard and Poors 500 companies providing environmental information in their annual financial reports rose from 217 in 1988 to 322 in 1990 (Ref. 24). With increasingly large environmental liability costs becoming a determinant of a corporate's financial success, there has also been increased pressure to monitor the adequacy of registrant's disclosure (Ref. 25).[39]

A 1992 Price Waterhouse survey indicated that the timing of the recording of the remediation (environmental) liabilities into the company's financial report varies. Some companies recorded different events at more than one stage. The timing of the recording in the financial report, for the companies responding to the survey, is as follows (Ref. 1):

- Upon internal discovery   56%
- Upon consent to remedial investigation and feasibility study   16%
  (RI/FS)
- During RI/FS   52%
- Upon completion of RI/FS   28%
- Upon settlement offer by the company   20%
- During cleanup   15%
- Upon notification by the regulatory authority   22%

The SEC enforcement works in collaboration with the EPA's enforcement division. The EPA maintains a current list of companies that are potentially responsible parties for those sites listed on the Superfund National Priority List. If the companies are on the National Priority List, but do not indicate

---

[39] The term "registrant" is used to describe companies registering public offerings of securities with the SEC under the Securities Act of 1933, 15 U.S.C. §§ 77a–77vvv, or filing periodic reports with the commission under the Securities and Exchange Act of 1934, 15 U.S.C. §§ 78–111.

that they are a potentially responsible party on their annual financial report, the SEC will follow up with inquiries. Such failure to disclose can result in fines and/or other penalties. When the SEC's Division of Corporate Finance staff find material omissions or deficiencies relating to environmental matters, it will request corrective disclosure and, in egregious cases, refer the matter to the SEC's Division of Enforcement who will then work with the EPA for appropriate enforcement actions. The SEC and the EPA have issued specific environmental disclosure requests to make the details about environmental programs and liabilities a matter of public record (Ref. 24).

Recently, the SEC commissioner stated that the SEC's role is that of both a regulator of corporate environmental conduct and an advocate for all investors who require information on environmental liabilities to make an informed investment decision. The industries specifically targeted for more intense scrutiny because they are more prone to environmental problems are the following:

- Pulp and paper companies
- Primary metal manufacturers
- Industrial organic chemical manufacturers

The EPA shares data with the SEC not only on Superfund sites, but also on toxic releases. The SEC compares the company's financial report disclosures with the EPA's information. Most of the work on specific accounting issues with the disclosure of environmental liabilities has been with Superfund remediation liabilities and potential responsible parties. Now there is a new interest to incorporate the proactive environmental auditing information with the accounting practices.

The American Institute of Certified Public Accountant's Environmental Accounting Task Force is developing accounting guidance in the recognition, measurement, display, and disclosure of environmental liabilities. The American Society for Testing and Materials, through its Environmental Risk Management Subcommittee on Reserve, is in the process of developing a voluntary standard "to define good commercial and customary practice in the United States for financial state disclosures regarding environmental impairment and associated liability." The objectives of the standard are to determine the conditions that warrant disclosure and to stress the appropriate disclosure requirements consistent with the level of environmental impairment. The GEMI has also just released their own guidance on environmental disclosures titled *Public Environmental Reporting Initiative.* SEC guidance concerning the current requirement for disclosures are summarized in Table 10.1 (Ref. 24). With many agencies collaborating on environmental issues, the general advice to corporations is to keep their disclosures consistent and to use the available environmental disclosure guidance documents.

Table 10.1 Securities and Exchange Commission Required Disclosure

Regulation	Requirements
Staff Accounting Bulletin 92	• Contingent liabilities should not be discounted, unless consistent with EITF 93-5. • Statements related to loss contingencies should include interest rate used, expected five-year aggregate, expected five-year payments, and thereafter, reconciliation of undiscounted amounts to total presented on balance sheet. • Separate presentation of gross liability from related claim for recovery except where right of set off exists. • Accrual of environmental exit costs over the life of the asset. • If failure of other potential responsible parties is "reasonably possible," disclose. • Measurement of liabilities based on currently available facts, existing technology, and presently enacted laws. • Recognize an amount equal to the lower amount of the range, even if the upper limit is uncertain (consistent with FASB Interpretation No. 14 — "Reasonable Estimation of the Amount of a Loss"). • Include in the notes to the Financial Statements, statements on judgments and assumptions, uncertainties on insurance, cost-sharing arrangements, time frame for payments, etc.
Regulation S-K	• Item 101 — Description of Business: "Disclosure shall be made as to the material effects that compliance with environmental laws may have upon capital expenditures, earnings, and competitive position of the registrant and its subsidiaries. The registrant shall disclose any material estimated capital expenditures for environmental control facilities for the remainder of the current fiscal year and for such further periods as the registrant may deem material." • Item 103 — Legal Proceedings: Item 103 requires disclosure regarding not only pending proceedings, but also "any such proceedings known to be contemplated by governmental authorities," must disclose material administrative or judicial proceedings. • Item 303 — MD&A: disclose environmental matters which materially affect liquidity, capital resources and results of operations.

Table 10.1 Securities and Exchange Commission Required Disclosure (Continued)

Regulation	Requirements
FASB 5: Accounting for Contingencies	• Loss should be reported if probable and reasonably estimable.
EITF 93-5: Accounting for Environmental Liabilities	• Discounting is appropriate if liability is "fixed and reliably determinable."
EITF 90-8: Capitalization of Costs to Treat Environmental Contamination	• Generally, should be charged to expense.
	• Capitalize if extends life, mitigates damage, or is incurred in preparing for sale.

Acronyms: EITF = Emerging Issues Task Force; MD&A = Management Discussions and Analysis; FASB = Financial Accounting Standard Board.

SECURITIES AND EXCHANGE COMMISSION GUIDANCE

To better understand the information in Table 10.1, it would be best to understand some of the accounting practices and issues. The following discussion is based on the *Environmental Liability Disclosure and Staff Accounting Bulletin No. 92* (Ref. 25). In dealing with disclosure of environmental liability issues, honest disclosure requirements are implied not only in the general federal securities law antifraud provisions, but also by the mandates of Regulation S-K.[40]

Regulations on Environmental Disclosure

The general antifraud provisions of the securities laws impose liabilities on those persons who make false statements or deliberate omissions of facts in connection with the offering, purchase, or sale of securities. In certain cases these general antifraud provisions require disclosure to investors of the effect of environmental laws on the issuer of the securities.

In addition to the general antifraud provisions, there are three provisions of the Regulation S-K that have important significance to registrants subject to potential environmental liabilities and risks. Item 101 (17 CFR 229.101)

---

[40] The SEC's system centers around Regulation S-K, 17 CFR 229.10 through 229.915, which is a list of reporting requirements for registrants.

requires a registrant to provide a general description of their business and specific disclosure of the effects that compliance with all environmental laws may have on their capital expenditures, earnings, and competitive position. Item 103 (17 CFR 229.103) requires the disclosure of any pending legal proceedings, including specified proceedings arising under federal or state environmental laws, and any proceedings being contemplated by governmental authorities. Item 303 (17 CFR 229.303), Management's Discussion and Analysis (MD&A), requires publicly held companies to include in their SEC filings a section containing a discussion of the registrant's historical results of managing environmental liabilities and risks, and any probable necessary future disclosures.

The provisions of the MD&A, the heart of disclosure, places the responsibility to disclose any concerns on the corporate management. In a 1989 interpretive release, the SEC stated that the probable future disclosure is triggered by any known trends, demands, commitments, events, or uncertainties that are reasonably likely to have an effect on the registrant's production or financial condition. The purpose of the MD&A disclosure is to give investors a look at the company through management's eyes, and, with its related financial statements, it is the heart of a registrant's disclosure document. Thus Item 303 will compel management to disclose any known or acknowledgeable significant implications of environmental laws on future operations of their company.

Environmental Accounting

Environmental matters may also have accounting implications for registrants. However, registrants appear not to be as familiar with the environmental liability accounting requirements as they are with the disclosure requirements. The methods for disclosure of environmental liabilities filed with the SEC are incorporated in the current accounting literature on accounting principles. However, there seems to be a great diversity in the application of the generally accepted accounting principles to the area of environmental liabilities. The diversity of the application tends to obscure the magnitude of the environmental liabilities and the related disclosures. One example, the timeliness of reporting environmental liabilities, was observed in the Price Waterhouse survey where there was a great variance in the timing of recording environmental liabilities in the financial report (Ref. 1).

To improve the disclosure of, and the accounting for environmental liabilities, the SEC released Staff Accounting Bulletin No. 92 to set forth the SEC's interpretation of the application of the generally accepted accounting principles. The bulletin represents the interpretations and practices followed by the SEC's Division of Corporate Finance and the Office of the Chief Accountant in administering the disclosure requirements of the federal securities laws.

Understanding the accounting literature on environmental liabilities is pertinent to understanding the federal security disclosure laws. The Financial Accounting Standards Board (FASB) Statement 5 is the recognized authority on the subject of contingent liabilities.[41] The statement contains key guidance on when a registrant must recognize a loss contingency in its financial statements and what relevant disclosures must be made. A loss contingency is defined in Statement 5 as follows:

> An existing condition, situation, or set of circumstances involving uncertainty as to possible . . . loss . . . to an enterprise that will ultimately be resolved when one or more future events occur or fail to occur. Resolution of the uncertainty may confirm the . . . impairment of an asset or incurrence of a liability.

The three pertinent disclosures pertaining to environmental contingencies are as follows:

- Disclosure of the nature of an accrual
- Disclosure of the amount accrued
- Disclosure of reasonably possible losses in excess of the amount accrued

If a company cannot estimate the range of reasonably possible outcomes, it must acknowledgment that in its financial statements. All delays for reasons of uncertainty in the estimation of loss or for reasons of waiting for a single amount of the loss are to be avoided. A range is acceptable, and the best estimate from the range or the lower end of the range should be reported as the liability.

Another issue recently observed has been that, for financial statement presentation purposes, the registrants were reducing the total expected liability for certain contingencies to take into account anticipated recoveries from a third party (insurance companies or other principally responsible parties) and the inflation rate. This practice is known as "offsetting" or "netting." While a company may be able to estimate the total amount expected to be paid for remediation of a particular site, it frequently cannot reliably estimate when those expenditures will occur. For example, a company may be able to estimate its liability for a particular site to be $40 million, but it has to determine the amount of cash expenditures required in the first year, the second year, etc. Discounting requires that the amount of cash expenditures for each period be known, and this is usually not the case. The appropriateness of netting or offsetting was resolved by the FASB's Emerging

---

[41] A contingent liability, in an environmental context, is a liability where the outcome is uncertain and the cost or dollar amount that the liability represents is not fixed.

Issues Task Force (EITF).[42] The EITF's response was "An environmental liability should be evaluated independently from any potential claim for recovery . . . and that the loss arising from the recognition of an environmental liability should be reduced only when a claim for recovery is probable of realization."

In turn, the SEC's Division of Corporate Finance responded in Staff Accounting Bulletin 92 by addressing the three key issues:

- The manner in which a contingent liability and any related asset representing claims for recovery should be displayed in the financial statements (offsetting)
- The appropriate discount rate to be used for recognition of a contingent liability at its present value (discounting)
- The disclosures likely to be of particular significance to investors in their assessment of contingent liabilities (disclosures)

The SEC holds strongly to the position that contingent liabilities must be displayed in the balance sheet separately from, and regardless of, amounts claimed from a third party, because this is the most fair presentation of the potential consequences of the contingent claim. However, many registrants have historically been using offsetting as a type of "safe harbor" based on the claim that it is difficult to estimate the environmental liabilities and, as a tool to conceal management's estimation of the potential liabilities, for protection from third parties. The SEC considers the presentation of the gross amount of liability, rather than the net, most fairly presents the registrant's resources. Offsetting the may leave the investors ignorant of the magnitude of the company's liability and may lead them to be less rigorous in considering other legal ramifications that may effect their investment. The SEC, in FASB interpretation No. 10, considers the offsetting of assets and liabilities in the balance sheet as improper under the general principles of accounting (except where a right of setoff exists). This position was recently strengthened by FASB Interpretation No. 39, which indicates that the prohibition on the offsetting of assets and liabilities in the balance sheet should be applied more comprehensively than it has previously. The use of offset remains a major controversy between industry and the SEC.

The second issue was the ability to recognize an estimated liability at its present value, rather than at the gross amount expected to be paid, i.e. when an environmental cleanup is completed, because the ultimate settlement of the environmental liabilities may not occur for years. Discounting the liability to

---

[42] The FASB established the EITF in July 1984 to assist the FASB in the early identification of emerging issues affecting financial reporting and of problems in implementing authoritative pronouncements.

reflect the time value of money (inflation) may be significant to some registrants. The SEC provides that discounting an environmental liability for a specific cleanup site is appropriate only if the aggregate amount of the obligation and the amount and timing of the cash payments are fixed or reliably determinable for that site. The SEC limited the discount rate to one equal to the rate on risk-free monetary assets, i.e., assets backed or guaranteed by the United States, with a maturity (final dollar value) equal to the expected cash payments of the environmental liability. The SEC considers that the liability should be discounted at a rate necessary to arrive at the amount at which the liability theoretically could be settled if the transaction with a third party occurred today and not in the future.

The third issue was a primary objective to elicit and standardize meaningful information concerning environmental matters in disclosure filings with the SEC. The disclosures should contain more specific and concrete information enabling investors to evaluate the nature and scope of the contingency on an informed basis. Too often, the disclosure is overly general and does not divulge sufficient information for the investor to speculate on the likely consequences.

All registrants should accrue a liability for the amounts they expect to pay, without regard to potential recoveries from third parties. SEC has identified those factors that registrants should consider in estimating the amount of a liability in Bulletin 92. The estimate should include the following:

- The estimate should be based on currently enacted laws and regulations, and on existing technology, rather than on assumption that environmental laws will change or that a new technology will enable the registrant to reduce the ultimate liability cost.
- Registrants should consider the effects of inflation and specific cost trends from prior industry experience in remediation released by regulatory agencies.

Bulletin 92 also provides examples of disclosures to be considered and emphasizes the disclosure obligations under Items 101, 103, and 303 of Regulation S-K. The SEC will be seeking disclosure of the range of reasonably possible outcomes in excess of the amount of the liability unless the registrants state that it is immaterial or it cannot be determined.

CONCLUSION

The full disclosure system requires disclosure of all known factors that would influence an investor's decision to buy or sell securities. The federal securities laws are designed to promote the full and fair disclosure of material facts. Registrants are encouraged to stay away from boilerplate, precautionary qualifying language that would make the disclosure ambiguous and to use "plain English" discussions of the significant factors that could influence the magnitude of the liability and the potential consequences to the registrant.[43]

Concerns with the environment have captured the public's attention, and environmental awareness probably is at an all-time high. This public interest has increased pressure on the SEC to ensure that registrants disclose in a fair, full, and timely manner present and potential future environmental costs of a material nature. The SEC's efforts and the need for better disclosure are expected to increase in the foreseeable future. The SEC's March 1992 action against Caterpillar, Inc. (when it failed to discuss uncertainties in the risk of a subsidiary's lower earnings) reaffirmed by its 1994 enforcement action against Shared Medical Systems Corp. (which failed to discuss that it was expecting a slowdown in growth due to declining sales) among others, should have delivered the message that the SEC considers MD&A disclosure to be a serious matter. As more data are accumulated and environmental litigation patterns become more definable, accounting for and disclosure of environmental liabilities should substantially improve.

---

[43] For example, a statement that the contingency is not expected to have a material effect on financial condition could be incomplete or confusing if the possible loss would be material to an investor based on another reasonable measure, such as one relating to liquidity or operating results. Further, this representation implies that management has determined the range of possible loss. If it is reasonably possible that the outcome of uncertainties may result in a liability materially exceeding the accrued liability, Statement 5 requires disclosure of that range of reasonably possible loss or a clear statement that a range cannot be estimated.

11   INTERNATIONAL AUDITS

INTERNATIONAL REGULATIONS AND RESOURCES

International harmonization of environmental standards and practices has been initiated with the ISO 14000 standards and will continue with the globalization of business practices. Global environmental regulations, treaties, and agreements are not new. The transborder Migratory Bird Treaty resulted in the protection of the Osprey and the transborder treaty on air pollution in Western Europe (1979) resulted in a dramatic fall in sulphur dioxide emissions, a precursor of acid rain. The Antarctic Treaty System (1991) banning mining exploration and development for 50 years, has protected the Antarctic from mining, military activities, nuclear tests, and radioactive wastes. The global ban on chlorofluorocarbons (related to ozone depletion) decreased consumption from 1.2 billion kilograms in 1987 to 682 million kilograms in 1991 (Ref. 26). The most recent and far-reaching agreement was the Rio Declaration adopted at the United Nations Conference on Environmental Development held in Brazil in 1992. The Rio Declaration outlines the interaction between independent states attaching greater importance to the social environment rather than the physical one. The five declaration principles outlining the most important principles of sustainable development are as follows (Ref. 27):[44]

- Emphasis on the quality of life rather than production
- A broadened concept integrating pollution and natural resources with the whole natural environment
- Treating the environment as an economic resource
- Differentiating between developed and developing countries' responsibilities for global environmental damage and response measures to potentially damaging activities
- Concern with broader national policies, strategies, and practices

---

[44] The Rio Declaration is the result of the United Nations Conference on Environmental Development held in Brazil in 1992. The Rio Declaration was adopted by more than 175 countries and the actual text of the Rio Declaration can be found at Rio Declaration on Environment and Development, United Nations Conference on Environment and Development, U.N. Doc. A/CONF.151/5/Rev. 1 (1992).

With the greater movement for international awareness and harmonization, it will be important to have a proactive auditing program that addresses the global practices issues and host country responsibilities and issues. Obviously, many countries are at different stages of environmental, safety, and health awareness and practices. Regulations in developing countries are influenced by the EPA and the World Bank Group; in the United Kingdom, Canada, and Japan by the Rio Declaration and the United Nations Conference on Environmental Development Agenda 21. With growing overseas operation, it becomes important not only to maintain a consistency of corporate environmental performance goals, but also to be in synchronization with host country concerns to meet the local regulations and requirements and to fit into the local culture.

In the beginning of this book, the usage and definition of the term "audit" was discussed as practiced in the United States. In foreign countries the term "audit" will have different culturally dependent connotations than it does in the United States. Some consider an audit to include environmental site assessments along with the standard usage of the term audit. Therefore it will be important to define the purpose and goal of the audit and to state what the audit will include. This may make the critical difference in how the audit proceeds.

There is no appreciable compilation of current foreign regulatory information translated into English. There are discrete translated versions available from some countries. For example, the *Environmental Acts of Korea,* published by the Ministry of Environment in Korea, is current on the major environmental acts of the country and useful to many United States auditors. Until more English translations of each country's regulations are available, the World Bank Group has some internal environmental guidance documents that may have influenced the local regulations and that may serve as an introductory reference. The World Bank Group also utilizes the United Nation Environment Program's *Guidelines for Environmental Auditing.*

In response to a growing need for auditing expertise and experience abroad, the EAR has initiated a database of auditors with international experience entitled *Member List by Country Experience.* The International Issues Work Group of the EAR has formed an International Regulatory Matrix Subcommittee. The initial intent was to supply members with a resource list and regulatory information in English where available. This task proved to be too difficult and the data always obsolete in an environment where the information is changing daily. Instead, a very successful attempt was made to identify people who had past international auditing experience and people who were planning international audits in the coming year. Members on the list are receptive to inquiries about their experience, and they may be the best and most current resource on local regulations. The latest issues of this list can be received by contacting the EAR Administrative Manager at telephone number 216-327-6605 or Fax 216-327-6609.

COMMENTS ON INTERNATIONAL AUDITING[45]

Environmental audits at international locations have a great many similarities to environmental audits in the continental United States. If the international components of a multinational company are expected to achieve the same basic environmental objectives as their United States facilities, then the audit program should be structured to audit all operations using the same basic procedures. It is important to maintain consistency so that the audit findings will be comparable. Auditors need to follow the same basic procedures; however, the successful performance of international audits requires certain adjustments to United States–oriented procedures. The adjustments must take into account regulatory and cultural differences, difficulties in communication, and a variety of logistical and practical challenges for auditors in a foreign country.

Regulatory Differences — Regulatory differences between the United States and other countries can be quite substantial; however, there are certainly more similarities than differences as they are supported by the same scientific principles. Similar environmental quality goals are promoted by international organizations, and the free flow of information among scientists and regulators around the world tends to create many regulatory similarities from country to country. There are differences though, and if an objective of the audit is to verify compliance with a nation's regulatory requirements, then the audit team must be properly trained in the national regulations.

A common approach is to have at least one team member who is familiar with both the host country's requirements and their implementation. United States team members may be brought up to speed on the applicable regulations of the host country. Several information systems firms offer at least substantial portions of many countries' health, safety, and environmental regulations. A training session for the audit team by an expert familiar with both the regulations and their enforcement can be helpful. Familiarity with the regulation should be supplemented by an understanding of the regulatory climate (culture). Developing countries in particular may not enforce some or all of their regulations. Learning local practices in advance from an independent source is obviously more desirable than soliciting the same information from the hosts.

Communications — Since the entire United States audit team is not likely to be proficient in the language of the host country, special steps will be needed to achieve an adequate level of communication. Some options are as follows:

- Engage a translator.
- Hire one or more native auditors proficient in the technical and regulatory issues to be covered.
- Employ an American bilingual auditor.

---

[45] This section of Chapter 11 was written by Mr. Ralph Rhodes.

Of the three, employing at least one local auditor has had the most satisfactory results. Familiarity with the practical aspects of regulation enforcement and knowledge of the local customs and language prepare the local auditor to contribute substantially to the work of the team. There should be a specific orientation for the local auditors to ensure that they are familiar with the team's methods, procedures, objectives, etc.

In situations where key plant personnel do not speak English and team members do not all speak the local language, it is extremely important that special steps be taken to assure full communication throughout the audit. Information gathering requires clear communication, and feedback to plant personnel is just as important. The English-speaking team members could, in the interest of saving time, skip the all-important feedback step, resulting in escalating tension and alienation among plant personnel. Frequent feedback sessions in the local language throughout the audit are important features. Translation takes time, and as much as 25% additional time could be required for an audit if most of the audit team cannot communicate directly with the plant personnel. The communication "surcharge" should be factored into the audit plan based on the language skills of plant personnel and the audit team.

Another communication challenge is that many international operations are unfamiliar with the Corporation's worldwide ES&H programs in general and the audit program in particular. Distance, culture, and language factors may combine to leave local management in the dark as to what the team expects to do and how they are expected to participate. The audit team leader should try to brief key plant personnel explicitly on the audit purpose, scope, and basic procedures to be followed. Where language difficulties occur, the information should be related through a liaison translator.

Cultural Differences — Many books have been written about cultural differences between the United States and other countries, particularly Asian countries. To the extent possible, the United States auditors should familiarize themselves with the cultural differences that could impact the satisfactory completion of the audit. To achieve success the key cultural differences must be discussed during the preaudit planning activities for the audit team.

For example, one major difference between United States and Asian countries in particular is the differing opportunity for women to advance to senior management positions. As a result, women auditors may find that answers to their questions are addressed to male members of the team. Women should not take offense at this practice, which is cultural and not personal, and should try to work around it.

A second cultural consideration is the custom in some countries to take long lunches and to entertain visitors generously. Pleasant as these social events are, they frequently intrude on audit work time. In these situations the team leader must balance team productivity against the objective of being a good guest in an unfamiliar land. A diplomatic explanation of the problem to the senior plant representative may result in less disruptions of the work schedule.

A third cultural consideration is the restraint of free communication on the part of junior-level staff members. In Asian cultures the society is organized and orderly. The young generation is taught to defer to their elders, as a matter of respect, and may not be as forthright in providing information. Therefore the auditors may need to initiate more response-provoking communications.

Logistical and Practical Considerations — There are obstacles that audit teams in foreign countries will have to overcome, particularly in developing countries. Communications to the corporate office and home may require calls at odd hours to allow for time differences. Sometimes the telephone system will not be working. Team members should advise family and friends that they may not be able to call regularly.

Travel requires more than a call to a car rental agency, particularly in some developing countries. Frequently, arranging for a local driver produces a better result at lower cost than using the normal routine followed in the United States.

The health of the team is a very important consideration. The team should avoid questionable public water supplies and exotic, native restaurants in developing countries. This can mean the difference between a team that is operational and one that is not.

Personal safety is a factor in some countries. Prudent steps should be taken to protect valuable equipment, such as laptop computers. Personal valuables should only be taken along when absolutely necessary. Corporate security departments often have useful advice for infrequent international visitors.

All of these considerations, and many more, could adversely affect the efficiency and effectiveness of the audit team assigned to an audit at an international location.

Prudent auditors will consult people who have traveled to the location for an audit and will solicit their advice. Local management should also be consulted. As hosts they will be anxious for the visit to be pleasant and for the audit to proceed smoothly. They will provide advice and frequently will make special arrangements that are difficult to make from the United States.

Above all, the team leader and the team should remember that they cannot assume that anything is the way it is in the United States. A combination of prior planning, flexibility in carrying out the plan, and prudence will provide the keys to avoiding logistical and practical difficulties.

REFERENCES

1. Stewart, Stephanie, The Voluntary Environmental Audit Survey of United States Business, Price Waterhouse L.L.P., presented at the Environmental Auditing Roundtable Conference, May 3, 1995.
2. Hayle, Brian and the Office of General Counsel, Chemical Manufacturers Association's *Environmental Audit Primer*, Chemical Manufacturers Association, Washington, DC, July 1992.
3. Environmental Protection Agency Final Auditing Policy: Environmental Auditing Policy Statement, *Federal Register* July 9, 1986 (51 FR 131, page 25004).
4. Environmental Protection Agency Final Auditing Policy: Incentives for Self-Policing: Discovery, Disclosure, Correction and Prevention, *Federal Register,* December 22, 1995 (60 FR 245, page 66706).
5. Karlin, Alex S., Conducting a Legal Checkup of an Environmental Audit Program, *Los Angeles Lawyer*, June 1994.
6. Markus, Kent, Acting Assistant Attorney General, United States Department of Justice letter to The Honorable Newt Gingrich, Speaker, United States House of Representatives, March 3, 1995.
7. Schiffer, Lois J., Assistant Attorney General, United States Department of Justice letter to Mr. Robert L Deschamps, President of National District Attorneys Association, May 15, 1995.
8. Baverstock, Suzie J., EARA Secretary and Executive Director of the Institute of Environmental Assessment, An Update on BS7750, and the Environmental Auditors Registration Scheme (EARA), presented at the Environmental Auditing Roundtable Meeting, April 1994.
9. Global Environmental Management Initiative Coalition. Environmental Self-Assessment Program, Global Environmental Management Initiative, Washington, DC, 1993.
10. Environmental Auditing Roundtable, Comments of the Environmental Auditing Roundtable, presented at Environmental Protection Agency Public Meeting on Audit Policy, July 27–28, 1994, Washington, DC.
11. Wallach, Paul G. and Anders, Sloane E., Statement of the Corporate Environmental Enforcement Council on Environmental Auditing Policy and Related Issues, presented at the Environmental Protection Agency Public Meeting on Audit Policy, July 27–28, 1994, Washington, DC.

12. Frick, G. William, Post-Meeting Statement of the American Petroleum Institute, Regarding the Environmental Protection Agency, July 27–28, 1994, Public Meeting on Environmental Auditing Policy and Related Environmental Compliance, Self-Evaluation and Disclosure Issues, August 11, 1994.
13. American Petroleum Institute Legal Counsel, Comments of the American Petroleum Institute, Regarding Interim Policy Statement on Voluntary Environmental Self-Policing and Self-Disclosure, 60 *Federal Register* 16875, September 1995.
14. Office of Management Data Systems, United States Department of Labor, Occupational Safety and Health Administration, Federal Standards Cited by Frequency for Fiscal Year 1994, Washington, DC.
15. Sarvadi, David G., The Egregious Penalty Policy and Violations of Specific Standards, *Compliance Magazine,* February 1996.
16. Halland, Elizabeth A. and Kroger, Elizabeth Mata, *The New, More Powerful OSHA*, Bracewell and Patterson, L.L.P., September 1994.
17. Sarvadi, David G., Workplace Health and Safety Regulation after 25 Years, *Journal of Environmental Law and Practice,* May/June 1995.
18. Segal, Scott H., *Occupational Safety and Health Reforms: "The New, More Powerful OSHA,"* Bracewell and Patterson, L.L.P., September 1994.
19. *OSHA Reform Faces Tough Road Ahead*, ENR, McGraw-Hill, New York, March 4, 1996.
20. American Industrial Hygiene Association Management Committee, *Industrial Hygiene Auditing — Manual for Practice*, American Industrial Hygiene Association Publication, Fairfax, VA, 1994.
21. McGuinness, Barbara J., DuPont's Environmental Audit Program, presented at Environmental Auditing Roundtable Conference, September 1995.
22. McCreary, Jean H., Implementing ISO 14000: Achieving Competitive Advantage Through Environmental Management, *Journal of Environmental Law and Practice,* 1994, Boston, MA.
23. Cahill, Lawrence B. and Schomer, Dawne P., The Potential Effect of ISO 14000 Standards on Environmental Audit Training in the United States, presented at the Environmental Auditing Roundtable Conference, January 1996.
24. Russel, William G., Proactive Environmental Accounting and the Annual Report, presented at the Environmental Auditing Roundtable Meeting, September 27, 1994, Baltimore, MD.
25. Environmental Liability Disclosure and Staff Accounting Bulletin No. 92, by Richard Y. Roberts, Commissioner, United States Securities and Exchange Commission, and Kurt R. Hohl, Associate Chief Accountant, Division of Corporate Finance United States Securities and Exchange Commission, *The Business Lawyer,* 50(1), November 1994, University of Maryland School of Law.
26. French, Hilary F., After the Earth Summit: The Future of Environmental Governance, Worldwatch Paper 107, Worldwatch Institute, Washington, DC, March 1992.
27. Kovat, Jeffrey D., A Short Guide to the Rio Declaration, *Colorado Journal International Environmental Policy,* 4, 119, 1993.

Appendix A
ENVIRONMENTAL PROTECTION AGENCY POLICY

Environmental Auditing Policy Statement 1986 140

Clarification on Policies Related to Environmental Auditing 157

Incentives for Self-Policing: Discovery, Disclosure,
Correction, and Prevention of Violations 1995 170

Federal Register / Vol. 51, No. 131 / Wednesday, July 9, 1986 / Notices

ENVIRONMENTAL AUDITING POLICY STATEMENT

Agency: Environmental Protection Agency (EPA).
Action: Final Policy Statement.

Summary: It is EPA policy to encourage the use of environmental auditing by regulated entities to help achieve and maintain compliance with environmental laws and regulations, as well as to help identify and correct unregulated environmental hazards. EPA first published this policy as interim guidance on November 8, 1985 (50 FR. 46504). Based on comments received regarding the interim guidance, the agency is issuing today's final policy statement with only minor changes. This final policy statement specifically does the following:

- Encourages regulated entities to develop, implement, and upgrade environmental auditing programs
- Discusses when the agency may or may not request audit reports
- Explains how EPA's inspection and enforcement activities may respond to regulated entities' efforts to ensure compliance through auditing
- Endorses environmental auditing at federal facilities
- Encourages state and local environmental auditing initiatives
- Outlines elements of effective audit programs

Environmental auditing includes a variety of compliance assessment techniques that go beyond those legally required and are used to identify actual and potential environmental problems. Effective environmental auditing can lead to higher levels of overall compliance and reduced risk to human health and the environment. EPA endorses the practice of environmental auditing and supports its accelerated use by regulated entities to help meet the goals of federal, state, and local environmental requirements. However, the existence of an auditing program does not create any defense for, or otherwise limit, the responsibility of any regulated entity to comply with applicable regulatory requirements.

States are encouraged to adopt these or similar and equally effective policies to advance the use of environmental auditing on a consistent, nationwide basis.

Dates: This final policy statement is effective July 9, 1986.

For further information contact: Leonard Fleckenstein, Office of Policy, Planning and Evaluation, (202) 382-2726; or Cheryl Wasserman, Office of Enforcement and Compliance Monitoring, (202) 382-7550.

SUPPLEMENTARY INFORMATION: ENVIRONMENTAL AUDITING POLICY STATEMENT

I. Preamble

On November 8, 1985, EPA published an Environmental Auditing Policy Statement, effective as interim guidance, and solicited written comments until January 7, 1986.

Thirteen commenters submitted written comments. Eight were from private industry. Two commenters represented industry trade associations. One federal agency, one consulting firm, and one law firm also submitted comments.

Twelve commenters addressed EPA requests for audit reports. Three comments per subject were received regarding inspections, enforcement response, and elements of effective environmental auditing. One commenter addressed audit provisions as remedies in enforcement actions, one addressed environmental auditing at federal facilities, and one addressed the relationship of the policy statement to state or local regulatory agencies. Comments generally supported both the concept of a policy statement and the interim guidance, but raised specific concerns with respect to particular language and policy issues in sections of the guidance.

General Comments

Three commenters found the interim guidance to be constructive, balanced, and effective at encouraging more and better environmental auditing.

Another commenter, while considering the policy on the whole to be constructive, felt that new and identifiable auditing "incentives" should be offered by EPA. Based on earlier comments received from industry, EPA believes most companies would not support or participate in an "incentives-based" environmental auditing programs with EPA. Moreover, general promises to forgo inspections or reduce enforcement responses in exchange for companies' adoption of environmental auditing program — the "incentive" most frequently mentioned in this context — are fraught with legal and policy obstacles.

APPENDIX A

Several commenters expressed concern that states or localities might use the interim guidance to require auditing. The agency disagrees that the policy statement opens the way for states and localities to require auditing.

No EPA policy can grant states or localities any more (or less) authority than they already possess. EPA believes that the interim guidance effectively encourages *voluntary* auditing. In fact, Section II.B of the policy states: "Because audit quality depends to a large degree on genuine management commitment to the program and its objectives, auditing should remain a voluntary program."

Another commenter suggested that EPA should not expect an audit to identify all potential problem areas or conclude that a problem identified in an audit reflects normal operations and procedures. EPA agrees that an audit report should clearly reflect these realities and should be written to point out the audit's limitations. However, since EPA will not routinely request audit reports, the agency does not believe these concerns raise issues that need to be addressed in the policy statement.

A second concern expressed by the same commenter was that EPA should acknowledge that environmental audits are only part of a successful environmental management program and thus should not be expected to cover every environmental issue or solve all problems. EPA agrees and accordingly has amended the statement of purpose, which appears at the end of this preamble.

Yet another commenter thought EPA should focus on environmental performance results (compliance or noncompliance), not on the processes or vehicles used to achieve those results. In general, EPA agrees with this statement and will continue to focus on environmental results. However, EPA also believes that such results can be improved through agency efforts to identify and encourage effective environmental management practices, and will continue to encourage such practices in nonregulatory ways.

A final general comment recommended that EPA should sponsor seminars for small businesses on how to start auditing programs. EPA agrees that such seminars would be useful. However, since audit seminars already are available from several private sector organizations, EPA does not believe it should intervene in that market, with the possible exception of seminars for government agencies, especially federal agencies, for which EPA has a broad mandate under Executive Order 12088 to provide technical assistance for environmental compliance.

Request for Reports

EPA received 12 comments regarding agency requests for environmental audit reports, far more than on any other topic in the policy statement. One commenter felt that EPA struck an appropriate balance between respecting the need for self-evaluation with some measure of privacy and allowing the agency enough flexibility of inquiry to accomplish future statutory missions. However, most commenters expressed concern that the interim guidance did not go far enough to assuage corporate fears that EPA will use audit reports for environmental compliance "witch hunts." Several commenters suggested additional specific assurances regarding the circumstances under which EPA will request such reports.

One commenter recommended that EPA request audit reports only "when the agency can show the information it needs to perform its statutory mission cannot be obtained from the monitoring, compliance, or other data that are otherwise reportable and/or accessible to EPA, or where the government deems an audit report material to a criminal investigation." EPA accepts this recommendation in part. The agency believes it would not be in the best interest of human health and the environment to commit to making a "showing" of a compelling information need before ever requesting an audit report. While EPA may normally be willing to do so, the agency cannot rule out in advance all circumstances in which such a showing may not be possible. However, it would be helpful to further clarify that a request for an audit report or a portion of a report normally will be made when needed information is not available by alternative means. Therefore EPA has revised Section III.A, paragraph two and added the phrase: "and usually made where the information needed cannot be obtained from monitoring, reporting, or other data otherwise available to the agency."

Another commenter suggested that (except in the case of criminal investigations) EPA should limit requests for audit documents to specific questions. By including the phrase "or relevant portions of a report" in Section III.A, EPA meant to emphasize it would not request an entire audit document when only a relevant portion would suffice. Likewise, EPA fully intends not to request even a portion of a report if needed information or data can be otherwise obtained. To further clarify this point EPA has added the phrase, "most likely focused on particular information needs rather than the entire report," to the second sentence of paragraph two, Section III.A. Incorporating the two comments above, the first two sentences in paragraph two of final Section III.A now read: "EPA's authority to request an audit report, or relevant portions thereof will be exercised on a case-by-case basis where the agency determines it is needed to accomplish a statutory mission or the government deems it to be material to a criminal investigation. EPA expects such requests to be limited, most likely focused on particular information needs rather than the entire report, and usually made where the information needed cannot be obtained from monitoring, reporting or other data otherwise available to the agency."

Other commenters recommended that EPA not request audit reports under any circumstances, that requests be "restricted to only those legally required," that requests be limited to criminal investigations, or that requests be made only when EPA has reason to believe "that the audit programs or reports are being used to conceal evidence of environmental noncompliance or otherwise being used in bad faith." EPA appreciates concerns underlying all of these comments and has considered each carefully. However, the agency believes that these recommendations do not strike the appropriate balance between retaining the flexibility to accomplish EPA's statutory missions in future, unforeseen circumstances, and acknowledging regulated entities' need to self-evaluate environmental performance with some measure of privacy. Indeed, based on prime informal comments, the small number of formal comments received, and the even smaller number of adverse comments, EPA believes the final policy statement should remain largely unchanged from the interim version.

Elements of Effective Environmental Auditing

Three commenters expressed concerns regarding the seven general elements EPA outlined in the Appendix to the interim guidance.

One commenter noted that were EPA to further expand or more fully detail such elements, programs not specifically fulfilling each element would then be judged inadequate. EPA agrees that presenting highly specific and prescriptive auditing elements could be counter-productive by not taking into account numerous factors that vary extensively from one organization to another, but that may still result in effective auditing programs. Accordingly, EPA does not plan to expand or more fully detail these auditing elements.

Another commenter asserted that states and localities should be cautioned not to consider EPA's auditing elements as mandatory steps. The agency is fully aware of this concern and in the interim guidance noted its strong opinion that "regulatory agencies should not attempt to prescribe the precise form and structure of regulated entities' environmental management or auditing programs." While EPA cannot require state or local regulators to adopt this or similar policies, the agency does strongly encourage them to do so, both in the interim and final policies.

A final commenter thought the Appendix too specifically prescribed what should and what should not be included in an auditing program. Other commenters, on the other hand, viewed the elements described as very general in nature. EPA agrees with these other commenters. The elements are in no way binding. Moreover, EPA believes that most mature, effective environmental auditing programs do incorporate each of these general elements in some form, and considers them useful yardsticks for those considering adopting or upgrading audit programs. For these reasons, EPA has not revised the Appendix in today's final policy statement.

Other Comments

Other significant comments addressed EPA inspection priorities for, and enforcement responses to, organizations with environmental auditing programs.

One commenter, stressing that audit programs are internal management tools, took exception to the phrase in the second paragraph of section III.B.1 of the interim guidance that states that environmental audits can "complement" regulatory oversight. By using the word "complement" in the context, EPA does not intend to imply that audit reports must be obtained by the agency to supplement regulatory inspection. "Complement" is used in a broad sense of being in addition to inspections and providing something (i.e., self-assessment) that otherwise would be lacking. To clarify this point EPA has added the phrase "by providing self-assessment to assure compliance" after "environmental audits may complement inspections" in this paragraph.

The same commenter also expressed concern that, as EPA sets inspection priorities, a company having an audit program could appear to be a "poor performer" due to complete and accurate reporting when measured against a company that reports something less than required by law. EPA agrees that it is important to communicate this fact to agency and state personnel and will do so. However, the agency does not believe a change in the policy statement is necessary.

A further comment suggested EPA should commit to take auditing programs into account when assessing all enforcement actions. However, to maintain enforcement flexibility under varied circumstances, the agency cannot promise reduced enforcement responses to violations at all audited facilities when other factors may be overriding. Therefore the policy statement continues to state that EPA may exercise its discretion to consider auditing programs as evidence of honest and genuine efforts to assure compliance, which would then be taken into account in fashioning enforcement responses to violations.

A final commenter suggested the phrase "expeditiously correct environmental problems" not be used in the enforcement context since it implied EPA would use an entity's record of correcting nonregulated matters when evaluating regulatory violations. EPA did not intend for such an inference to be made. EPA intended the term "environmental problems" to refer to the underlying circumstances that eventually lead up to the violations. To clarify this point, EPA is revising the first two sentences of the paragraph to which this comment refers by changing "environmental problems" to "violations and underlying environmental problems" in the first sentence and to "underlying environmental problems" in the second sentence.

In a separate development EPA is preparing an update of its January 1984 *Federal Facilities Compliance Strategy,* which is referenced in section III.C of the auditing policy. The Strategy should be completed and available on request from EPA's Office of Federal Activities later this year.

EPA thanks all commenters for responding to the November 8, 1985, publication. Today's notice is being issued to inform regulated entities and the public of EPA's final policy toward environmental auditing. This policy was developed to help (a) encourage regulated entities to institutionalize effective audit practices as one means of improving compliance and sound environmental management and (b) guide internal EPA actions directly related to regulated entities' environmental auditing programs.

EPA will evaluate implementation of this final policy to ensure it meets the above goals and continues to encourage better environmental management while strengthening the agency's own efforts to monitor and enforce compliance with environmental requirements.

II. General EPA Policy on Environmental Auditing

A. *Introduction*

Environmental auditing is a systematic, documented, periodic, and objective review by regulated entities[1] of facility operations and practices related to meeting environmental requirements. Audits can be designed to accomplish any or all of the following: verify compliance with environmental requirements; evaluate the effectiveness of environmental management systems already in place; or assess risks from regulated and unregulated materials and practices.

Auditing serves as a quality assurance check to help improve the effectiveness of basic environmental management by verifying that management practices are in place, functioning, and adequate. Environmental audits evaluate, and are not a substitute for, direct compliance activities such as obtaining permits, installing controls, monitoring compliance, reporting violations, and keeping records. Environmental auditing may verify, but does not include, activities required by law, regulation, or permit (e.g., continuous emissions monitoring, composite correction plans at wastewater treatment plants, etc.). Audits do not in any way replace regulatory agency inspections. However, environmental audits can improve compliance by complementing conventional federal, state, and local oversight.

The appendix to this policy statement outlines some basic elements of environmental auditing (e.g., auditor independence and top management support) for use by those considering implementation of effective auditing programs to help achieve and maintain compliance. Additional information on environmental auditing practices can be found in various published materials.[2]

Environmental auditing has developed for sound business reasons, particularly as a means of helping regulated entities manage pollution control affirmatively over time instead of reacting to crises. Auditing can result in improved

---

[1] "Regulated entities" include private firms and public agencies with facilities subject to environmental regulation. Public agencies can include federal, state, or local agencies, as well as special-purpose organizations such as regional sewage commissions.

[2] See, e.g., "Current Practices in Environmental Auditing," EPA Report No. EPA-230-09-83-006, February 1984: "Annotated Bibliography on Environmental Auditing," Fifth Edition, September 1985, both available from: Regulatory Reform Staff, PM-223, EPA, 401 M Street SW, Washington, DC 20460.

facility environmental performance, help communicate effective solutions to common environmental problems, focus facility managers' attention on current and upcoming regulatory requirements, and generate protocols and checklists that help facilities better manage themselves. Auditing also can result in better-integrated management of environmental hazards, since auditors frequently identify environmental liabilities that go beyond regulatory compliance. Companies, public entities, and federal facilities have employed a variety of environmental auditing practices in recent years. Several hundred major firms in diverse industries now have environmental auditing programs, although they often are known by other names, such as assessment, survey, surveillance, review, or appraisal.

While auditing has demonstrated its usefulness to those with audit programs, many others still do not audit. Clarification of EPA's position regarding auditing may help encourage regulated entities to establish audit programs or upgrade systems already in place.

B. EPA Encourages the Use of Environmental Auditing

EPA encourages regulated entities to adopt sound environmental management practices to improve environmental performance. In particular, EPA encourages regulated entities subject to environmental regulations to institute environmental auditing programs to help ensure the adequacy of internal systems to achieve, maintain, and monitor compliance. Implementation of environmental auditing programs can result in better identification, resolution, and avoidance of environmental problems, as well as improvements to management practices. Audits can be conducted effectively by independent internal or third-party auditors. Larger organizations generally have greater resources to devote to an internal audit team, while smaller entities might be more likely to use outside auditors.

Regulated entities are responsible for taking all necessary steps to ensure compliance with environmental requirements, whether or not they adopt audit programs. Although environmental laws do not require a regulated facility to have an auditing program, ultimate responsibility for the environmental performance of the facility lies with top management, which therefore has a strong incentive to use reasonable means, such as environmental auditing, to secure reliable information of facility compliance status.

EPA does not intend to dictate or interfere with the environmental management practices of private or public organizations. Nor does EPA intend to mandate auditing (though in certain instances EPA may seek to include provisions for environmental auditing as part of settlement agreements, as noted below). Because environmental auditing systems have been widely adopted on a voluntary basis in the past, and because audit quality depends to a large degree upon genuine management commitment to the program and its objectives, auditing should remain a voluntary activity.

III. EPA Policy on Specific Environmental Auditing Issues

A. Agency Requests for Audit Reports

EPA has broad statutory authority to request relevant information on the environmental compliance status of regulated entities. However, EPA believes routine agency requests for audit reports[3] could inhibit auditing in the long run, decreasing both the quantity and quality of audits conducted. Therefore, as a matter of policy, EPA will not routinely request environmental audit reports.

EPA's authority to request an audit report, or relevant portions thereof, will be exercised on a case-by-case basis where the agency determines it is needed to accomplish a statutory mission, or where the government deems it to be material to a criminal investigation. EPA expects such requests to be limited, most likely focused on particular information needs rather than on the entire report, and usually made where the information needed cannot be obtained from monitoring, reporting, or other data otherwise available to the agency. Examples would likely include situations where audits are conducted under consent decrees or other settlement agreements; a company has placed its management practices at issue by raising them as a defense; or state of mind or intent are a relevant element of inquiry, such as during a criminal investigation. This list is illustrative rather than exhaustive, since there doubtless will be other situations, not subject to prediction, in which audit reports rather than information may be required.

EPA acknowledges regulated entities' need to self-evaluate environmental performance with some measure of privacy and encourages such activity. However, audit reports may not shield monitoring, compliance, or other information that would otherwise be reportable and/or accessible to EPA, even if there is no explicit "requirement" to generate that data.[4] Thus this policy does not alter regulated entities' existing or future obligations to monitor, record, or report information required under environmental statutes, regulations, or permits, or to allow EPA access to that information. Nor does this policy alter EPA's authority to request and receive any relevant information — including that contained in audit reports — under various environmental statutes (e.g., Clean Water Act section 308, Clean Air Act sections 114 and 208) or in other administrative or judicial proceedings.

Regulated entities also should be aware that certain audit findings may by law have to be reported to government agencies. However, in addition to

---

[3] An "environmental audit report" is a written report that candidly and thoroughly presents findings from a review, conducted as part of an environmental audit as described in section II.A, of facility environmental performance and practices. An audit report is not a substitute for compliance monitoring reports or other reports or records that may be required by EPA or other regulatory agencies.

[4] See, for example, "Duties to Report or Disclose Information on the Environmental Aspects of Business Activities." Environmental Law Institute report to EPA, final report, September 1985.

any such requirements, EPA encourages regulated entities to notify appropriate state or federal officials of findings that suggest significant environmental or public health risks, even when not specifically required to do so.

B. EPA Response to Environmental Auditing

1. General Policy

EPA will not promise to forgo inspections, reduce enforcement responses, or offer other such incentives in exchange for implementation of environmental auditing or other sound environmental management practices. Indeed, a credible enforcement program provides a strong incentive for regulated entities to audit.

Regulatory agencies have an obligation to assess source compliance status independently and cannot eliminate inspections for particular firms or classes of firms. Although environmental audits may complement inspections by providing self-assessment to assure compliance, they are in no way a substitute for regulatory oversight. Moreover, certain statutes (e.g., RCRA) and agency policies establish minimum facility inspection frequencies to which EPA will adhere.

However, EPA will continue to address environmental problems on a priority basis and will consequently inspect facilities with poor environmental records and practices more frequently. Since effective environmental auditing helps management identify and promptly correct actual or potential problems, audited facilities' environmental performance should improve. Thus, while EPA inspections of self-audited facilities will continue, to the extent that compliance performance is considered in setting inspection priorities, facilities with a good compliance history may be subject to fewer inspections.

In fashioning enforcement responses to violations, EPA policy is to take into account, on a case-by-case basis, the honest and genuine efforts of regulated entities to avoid and promptly correct violations and underlying environmental problems. When regulated entities take reasonable precautions to avoid noncompliance, expeditiously correct underlying environmental problems discovered through audits or other means, and implement measures to prevent their recurrence, EPA may exercise its discretion to consider such actions as honest and genuine efforts to assure compliance. Such consideration applies particularly when a regulated entity promptly reports violations or compliance data that otherwise were not required to be recorded or reported to EPA.

2. Audit Provisions as Remedies in Enforcement Actions

EPA may propose environmental auditing provisions in consent decrees and in other settlement negotiations where auditing could provide a remedy for identified problems and reduce the likelihood of similar problems recurring in the future.[5] Environmental auditing provisions are most likely to be proposed in settlement negotiations where

- A pattern of violations can be attributed, at least in part, to the absence or poor functioning of an environmental management system; or
- The type or nature of violations indicates a likelihood that similar noncompliance problems may exist or occur elsewhere in the facility or at other facilities operated by the regulated entity.

Through this consent decree approach and other means, EPA may consider how to encourage effective auditing by publicly owned sewage treatment works (POTWs). POTWs often have compliance problems related to operation and maintenance procedures that can be addressed effectively through the use of environmental auditing. Under its National Municipal Policy EPA already is requiring many POTWs to develop composite correction plans to identify and correct compliance problems.

C. Environmental Auditing at Federal Facilities

EPA encourages all federal agencies subject to environmental laws and regulations to institute environmental auditing systems to help ensure the adequacy of internal systems to achieve, maintain, and monitor compliance. Environmental auditing at federal facilities can be an effective supplement to EPA and state inspections. Such federal facility environmental audit programs should be structured to promptly identify environmental problems and expeditiously develop schedules for remedial action.

To the extent feasible, EPA will provide technical assistance to help federal agencies design and initiate audit programs. Where appropriate, EPA will enter into agreements with other agencies to clarify the respective roles, responsibilities, and commitments of each agency in conducting and responding to federal facility environmental audits.

With respect to inspections of self-audited facilities (see section III.B.1 above) and requests for audit reports (see section III.A above), EPA generally will respond to environmental audits by federal facilities in the same manner as it does for other regulated entities, in keeping with the spirit and intent of Executive Order 12088 and the EPA *Federal Facilities Compliance Strategy* (January 1984, update forthcoming in late 1986). Federal agencies should, however, be aware that the Freedom of Information Act will govern any disclosure of audit reports or audit-generated information requested from federal agencies by the public.

---

[5] EPA is developing guidance for use by agency negotiators in structuring appropriate environmental audit provisions for consent decrees and other settlement negotiations.

When federal agencies discover significant violations through an environmental audit, EPA encourages them to submit the related audit findings and remedial action plans expeditiously to the applicable EPA regional office (and responsible state agencies, where appropriate) even when not specifically required to do so. EPA will review the audit findings and action plans and either provide written approval or negotiate a Federal Facilities Compliance Agreement. EPA will utilize the escalation procedures provided in Executive Order 12088 and the EPA *Federal Facilities Compliance Strategy* only when agreement between agencies cannot be reached. In any event, federal agencies are expected to report pollution abatement projects involving costs (necessary to correct problems discovered through the audit) to EPA in accordance with OMB Circular A-106. Upon request, EPA will assist affected federal agencies through coordination of any public release of audit findings with approved action plans once agreement has been reached.

IV. Relationship to State or Local Regulatory Agencies

State and local regulatory agencies have independent jurisdiction over regulated entities. EPA encourages them to adopt these or similar policies, in order to advance the use of effective environmental auditing in a consistent manner.

EPA recognizes that some states have already undertaken environmental auditing initiatives that differ somewhat from this policy. Other states also may want to develop auditing policies that accommodate their particular needs or circumstances. Nothing in this policy statement is intended to preempt or preclude states from developing other approaches to environmental auditing. EPA encourages state and local authorities to consider the basic principles that guided the agency in developing this policy:

- Regulated entities must continue to report or record compliance information required under existing statutes or regulations, regardless of whether such information is generated by an environmental audit or contained in an audit report. Required information cannot be withheld merely because it is generated by an audit rather than by some other means.
- Regulatory agencies cannot make promises to forgo or limit enforcement action against a particular facility or class of facilities in exchange for the use of environmental auditing systems. However, such agencies may use their discretion to adjust enforcement actions on a case-by-case basis in response to honest and genuine efforts by regulated entities to assure environmental compliance.
- When setting inspection priorities, regulatory agencies should focus to the extent possible on compliance performance and environmental results.

APPENDIX A 153

- Regulatory agencies must continue to meet minimum program requirements (e.g., minimum inspection requirements, etc.).
- Regulatory agencies should not attempt to prescribe the precise form and structure of regulated entities' environmental management or auditing programs.

An effective state/federal partnership is needed to accomplish the mutual goal of achieving and maintaining high levels of compliance with environmental laws and regulations. The greater the consistency between state or local policies and this federal response to environmental auditing, the greater the degree to which sound auditing practices might be adopted and compliance levels improve.

Dated: June 28, 1986
Lee M. Thomas, Administrator

Appendix—Elements of Effective Environmental Auditing Programs

Introduction: Environmental auditing is a systematic, documented, periodic, and objective review by a regulated entity of facility operations and practices related to meeting environmental requirements.

Private sector environmental audits of facilities have been conducted for several years and have taken a variety of forms, in part to accommodate unique organizational structures and circumstances. Nevertheless, effective environmental audits appear to have certain discernible elements in common with other kinds of audits. Standards for internal audits have been documented extensively. The elements outlined below draw heavily on two of these documents: "Compendium of Audit Standards" (©1983, Walter Willborn, American Society for Quality Control) and "Standards for the Professional Practice of Internal Auditing" (©1981, The Institute of Internal Auditors, Inc.). They also reflect agency analyses conducted over the last several years.

Performance-oriented auditing elements are outlined here to help accomplish several objectives. A general description of features of effective, mature audit programs can help those starting audit programs, especially federal agencies and smaller businesses. These elements also indicate the attributes of auditing EPA generally considers important to ensure program effectiveness. Regulatory agencies may use these elements in negotiating environmental auditing provisions for consent decrees. Finally, these elements can help guide states and localities considering auditing initiatives.

An effective environmental auditing system will likely include the following general elements:

I. *Explicit top management support for environmental auditing and commitment to follow up on audit findings.* Management support may be demonstrated by a written policy articulating upper management support for the auditing program and for compliance with all pertinent requirements, including corporate policies and permit requirements, as well as federal, state, and local statutes and regulations. Management support for the auditing program also should be demonstrated by an explicit written commitment to follow up on audit findings to correct identified problems and prevent their recurrence.

II. *An environmental auditing function independent of audited activities.* The status or organizational locus of environmental auditors should be sufficient to ensure objective and unobstructed inquiry, observation, and testing. Auditor objectivity should not be impaired by personal relationships, financial or other conflicts of interest, interference with free inquiry or judgment, or fear of potential retribution.

III. *Adequate team staffing and auditor training.* Environmental auditors should possess or have ready access to the knowledge, skills, and disciplines needed to accomplish audit objectives. Each individual auditor should comply with the company's professional standards of conduct. Auditors, whether full-time or part-time, should maintain their technical and analytical competence through continuing education and training.

IV. *Explicit audit program objectives, scope, resources, and frequency.* At a minimum, audit objectives should include assessing compliance with applicable environmental laws and evaluating the adequacy of internal compliance policies, procedures, and personnel training programs to ensure continued compliance.

Audits should be based on a process that provides auditors: all corporate policies, permits, and federal, state, and local regulations pertinent to the facility; and checklists or protocols addressing specific features that should be evaluated by auditors.

Explicit written audit procedures generally should be used for planning audits, establishing audit scope, examining and evaluating audit findings, communicating audit results, and following up.

V. *A process that collects, analyzes, interprets, and documents information sufficient to achieve audit objectives.* Information should be collected before and during an on-site visit regarding environmental compliance,[6] environmental management effectiveness,[7] and other matters[8] related to audit objectives and scope. This information should be sufficient, reliable, relevant, and useful to provide a sound basis for audit findings and recommendations.

 a. *Sufficient* information is factual, adequate, and convincing so that a prudent, informed person would be likely to reach the same conclusions as the auditor.

 b. *Reliable* information is the best attainable through use of appropriate audit techniques.

APPENDIX A 155

    c. *Relevant* information supports audit findings and recommendations and is consistent with the objectives for the audit.
    d. *Useful* information helps the organization meet its goals.
The audit process should include a periodic review of the reliability and integrity of this information and the means used to identify, measure, classify, and report it. Audit procedures, including the testing and sampling techniques employed, should be selected in

---

[6] A comprehensive assessment of compliance with federal environmental regulations requires an analysis of facility performance against numerous environmental statutes and implementing regulations. These statutes include: Resource Conservation and Recovery Act; Federal Water Pollution Control Act; Clean Air Act; Hazardous Materials Transportation Act; Toxic Substances Control Act; Comprehensive Environmental Response, Compensation and Liability Act; Safe Drinking Water Act; Federal Insecticide, Fungicide and Rodenticide Act; Marine Protection, Research and Sanctuaries Act; and Uranium Mill Tailings Radiation Control Act.

In addition, state and local government are likely to have their own environmental laws. Many states have been delegated authority to administer federal programs. Many local governments' building, fire, safety, and health codes also have environmental requirements relevant to an audit evaluation.

[7] An environmental audit could go well beyond the type of compliance assessment normally conducted during regulatory inspections, for example, by evaluating policies and practices, regardless of whether they are part of the environmental system or the operating and maintenance procedures. Specifically, audits can evaluate the extent to which systems or procedures:

1. Develop organizational environmental policies that (a) implement regulatory requirements and (b) provide management guidance for environmental hazards not specifically addressed in regulations.
2. Train and motivate facility personnel to work in an environmentally acceptable manner and to understand and comply with government regulations and the entity's environment policy.
3. Communicate relevant environmental developments expeditiously to facility and other personnel.
4. Communicate effectively with government and the public regarding serious environmental incidents.
5. Require third parties working for, with, or on behalf of the organization to follow its environmental procedures.
6. Make proficient personnel available at all times to carry out environmental (especially emergency) procedures.
7. Incorporate environmental protection into written operating procedures.
8. Apply the best management practices and operating procedures, including "good housekeeping" techniques.
9. Institute preventive and corrective maintenance systems to minimize actual and potential environmental harm.
10. Utilize the best available process and control technologies.
11. Use most effective sampling and monitoring techniques, test methods, recordkeeping systems, or reporting protocols (beyond minimum legal requirements).
12. Evaluate causes behind any serious environmental incidents and establish procedures to avoid recurrence.
13. Exploit source reduction, recycle, and reuse potential wherever practical.
14. Substitute materials or processes to allow use of the least hazardous substances feasible.

[8] Auditors could also assess environmental risks and uncertainties.

advance, to the extent practical, and expanded or altered if circumstances warrant. The process of collecting, analyzing, interpreting, and documenting information should provide reasonable assurance that audit objectivity is maintained and audit goals are met.

VI. *A process that includes specific procedures to promptly prepare candid, clear, and appropriate written reports on audit findings, corrective actions, and schedules for implementation.* Procedures should be in place to ensure that such information is communicated to managers, including facility and corporate management, who can evaluate the information and ensure correction of identified problems. Procedures also should be in place for determining what internal findings are reportable to state or federal agencies.

VII. *A process that includes quality assurance procedures to assure the accuracy and thoroughness of environmental audits.* Quality assurance may be accomplished through supervision, independent internal reviews, external reviews, or a combination of these approaches.

Federal Register / Vol. 59 / Thursday, July 28, 1994 / Notices

CLARIFICATION ON POLICIES RELATED TO ENVIRONMENTAL AUDITING

Agency: Environmental Protection Agency (EPA).
Action: Restatement of Policies Related to Environmental Auditing.

I. Auditing Public Meeting: Change of Location

The response to EPA's announcement (59 FR 31914, June 20, 1994) to hold a public meeting on auditing on July 27–28, 1994, has been overwhelming. Due to the expected size of the audience, therefore, the agency has changed the location of this event. The new location is the Stouffer Mayflower Hotel in Washington, DC, at 1127 Connecticut Avenue, NW, Phone (202) 347-3000.

II. The Auditing Policy Reassessment

In response to a request by Administrator Carol M. Browner, the Office of Enforcement and Compliance Assurance (OECA) is reassessing the agency's current policy regarding environmental auditing and self-evaluation by the regulated community. EPA has committed to investigating the perceived problems relating to auditing, self-evaluation, and disclosure through an empirical, information-gathering effort. The agency must develop an adequate information base to give serious consideration to any policy options and to ensure that any decision to either reinforce, change, or supplement existing policy is informed by fact.

EPA hopes to collect such relevant data through the implementation of four actions this summer. First, the agency will convene a public meeting on July 27–28, 1994, as an opportunity to obtain a wide variety of views and to sharpen the focus on these issues. The range of issues appropriate for discussion at the public meeting include the implementation of the 1986 policy; specific suggestions for auditing policy options; state audit privilege legislation; auditing in the context of criminal enforcement; and advances in the field of auditing since 1986. Interested parties are encouraged to read the *Federal Register* notice dated June 20, 1994 (59 FR 31914) for more details on the public meeting.

Second, EPA published in the June 21, 1994, *Federal Register* (59 FR 32062) a notice requesting proposals for Environmental Leadership Program (ELP) pilot projects. EPA expects that these pilot projects will generate useful data on auditing methodology and measures, and may also serve as vehicles for experimenting with policy-driven incentives.

Third, EPA will encourage the private sector to collect data and survey auditing practices in order to gauge the effect of enforcement policies on self-evaluation and disclosure in the regulated community. The agency will also seek input on auditing and related issues from states, environmental, and public interest groups, and trade and professional associations.

Finally, in this *Federal Register* notice, EPA is restating salient points from the 1986 policy and reviewing its activities and other policies relating to environmental auditing. The goal of this notice is to clarify EPA's current policies on and approach to auditing, in order to ensure a well-informed policy debate.

III. Review of General EPA Policy on Environmental Auditing

A. *EPA Encourages the Use of Environmental Auditing*

EPA has actively encouraged and participated in the development of environmental auditing and improved environmental management practices since the mid-1980s. In fact, the 1986 policy has served as the basis for defining the practice and profession of environmental auditing. The 1986 policy clearly states EPA support for auditing:

Effective environmental auditing can lead to higher levels of overall compliance and reduced risk to human health and the environment. EPA endorses the practice of environmental auditing and supports its accelerated use by regulated entities to help meet the goals of federal, state, and local environmental requirements.

Auditing serves as a quality assurance check to help improve the effectiveness of basic environmental management by verifying that management practices are in place, functioning, and adequate. Environmental audits evaluate, and are not a substitute for, direct compliance activities, such as obtaining permits, installing controls, monitoring compliance, reporting violations, and keeping records. Environmental auditing may verify, but does not include, activities required by law, regulation, or permit (e.g., continuous emissions monitoring, composite correction plans at wastewater treatment plants, etc.). Audits do not in any way replace regulatory agency inspections. However, environmental audits can improve compliance by complementing conventional federal, state, and local oversight.

Environmental auditing has developed for sound business reasons, particularly as a means of helping regulated entities manage pollution control affirmatively over time instead of reacting to crises. Auditing can result in improved facility environmental performance, help communicate effective solutions to common environmental problems, focus facility managers' attention on current and upcoming regulatory requirements, and generate protocols and checklists that help facilities better manage themselves. Auditing also can result in better integrated management of environmental hazards, since auditors frequently identify environmental liabilities that go beyond regulatory compliance.

APPENDIX A

The agency clearly supports auditing to help ensure the adequacy of internal systems to achieve, maintain, and monitor compliance. By voluntarily implementing environmental management and auditing programs, regulated entities can identify, resolve, and avoid environmental problems.

EPA does not intend to dictate or interfere with the environmental management practices of private or public organizations. Nor does EPA intend to mandate auditing (though in certain instances EPA may seek to include provisions for environmental auditing as part of settlement agreements, as noted below). Because environmental auditing systems have been widely adopted on a voluntary basis in the past, and because audit quality depends to a large degree upon genuine management commitment to the program and its objectives, auditing should remain a voluntary activity.

Because senior managers of regulated entities are ultimately responsible for taking all necessary steps to ensure compliance with environmental requirements, EPA believes they have a strong incentive to use reasonable means, such as environmental auditing, to secure reliable information about facility compliance status.

B. Definition of Environmental Auditing, Elements of Effective Environmental Auditing Programs

The 1986 policy also defines environmental auditing and outlines what EPA considers to be the elements of an effective environmental auditing program. The 1986 policy presents the following definition:

Environmental auditing is a systematic, documented, periodic, and objective review by regulated entities of facility operations and practices related to meeting environmental requirements. Audits can be designed to accomplish any or all of the following: verify compliance with environmental requirements; evaluate the effectiveness of environmental management systems already in place; or assess risks from regulated and unregulated materials and practices.

An organization's auditing program will evolve according to its unique structures and circumstances. The 1986 policy acknowledges this fact and also states EPA's belief that effective environmental auditing programs appear to have certain discernible elements in common with other kinds of audit programs. EPA generally considers these elements important to ensure program effectiveness. This general description of effective, mature audit programs can help those starting audit programs, especially federal agencies and smaller businesses. Regulatory agencies may also use these elements in negotiating environmental auditing provisions for consent decrees. Finally, these elements can help guide states and localities considering auditing initiatives.

As stated in the 1986 policy, an effective environmental auditing system will likely include the following general elements:

I. *Explicit top management support for environmental auditing and commitment to follow up on audit findings.* Management support may be demonstrated by a written policy articulating upper management support for the auditing program and for compliance with all pertinent requirements, including corporate policies and permit requirements, as well as federal, state, and local statutes and regulations.

Management support for the auditing program also should be demonstrated by an explicit written commitment to follow up on audit findings to correct identified problems and prevent their recurrence.

II. *An environmental auditing function independent of audited activities.* The status or organizational locus of environmental auditors should be sufficient to ensure objective and unobstructed inquiry, observation, and testing. Auditor objectivity should not be impaired by personal relationships, financial or other conflicts of interest, interference with free inquiry or judgment, or fear of potential retribution.

III. *Adequate team staffing and auditor training.* Environmental auditors should possess or have ready access to the knowledge, skills, and disciplines needed to accomplish audit objectives. Each individual auditor should comply with the company's professional standards of conduct. Auditors, whether full-time or part-time, should maintain their technical and analytical competence through continuing education and training.

IV. *Explicit audit program objectives, scope, resources, and frequency.* At a minimum, audit objectives should include assessing compliance with applicable environmental laws and evaluating the adequacy of internal compliance policies, procedures, and personnel training programs to ensure continued compliance.

Audits should be based on a process that provides auditors: all corporate policies, permits, and federal, state, and local regulations pertinent to the facility; and checklists or protocols addressing specific features that should be evaluated by auditors.

Explicit written audit procedures generally should be used for planning audits, establishing audit scope, establishing audit scope, examining and evaluating audit findings, communicating audit results, and following up.

V. *A process that collects, analyzes, interprets, and documents information sufficient to achieve audit objectives.* Information should be collected before and during an on-site visit regarding environmental compliance, environmental management effectiveness, and other matters related to audit objectives and scope. This information should be sufficient, reliable, relevant, and useful to provide a sound basis for audit finds and recommendations.

APPENDIX A 161

  a. *Sufficient* information is factual, adequate, and convincing so that a prudent, informed person would be likely to reach the same conclusions as the auditor.
  b. *Reliable* information is the best attainable through use of appropriate audit techniques.
  c. *Relevant* information supports audit findings and recommendations and is consistent with the objectives for the audit.
  d. *Useful* information helps the organization meet its goals.

  The audit process should include a periodic review of the reliability and integrity of this information and the means used to identify, measure, classify, and report it. Audit procedures, including the testing and sampling techniques employed, should be selected in advance, to the extent practical, and expanded or altered if circumstances warrant. The process of collecting, analyzing, interpreting and documenting information should provide reasonable assurance that audit objectivity is maintained and audit goals are met.

VI. *A process that includes specific procedures to promptly prepare candid, clear, and appropriate written reports on audit findings, corrective actions, and schedules for implementation.* Procedures should be in place to ensure that such information is communicated to managers, including facility and corporate management, who can evaluate the information and ensure correction of identified problems. Procedures also should be in place for determining what internal findings are reportable to state or federal agencies.

VII. *A process that includes quality assurance procedures to assure the accuracy and thoroughness of environmental audits.* Quality assurance may be accomplished through supervision, independent internal reviews, external reviews, or a combination of these approaches.

C. EPA Activities Related to Auditing Standards

EPA is currently participating in two major nonregulatory efforts to develop voluntary standards for auditing and environmental management systems. First, the International Organization of Standards (ISO), based in Geneva, Switzerland, established in 1993 a Technical Committee for Environmental Management Standards (ISO-TC-207). Subcommittee Two of TC-207 is in the process of developing environmental auditing standards. The standards fall into three groups: Auditing Principles, Auditing Procedures, and Auditor Qualifications. Second, in the United States, the National Sanitation Foundation (NSF) in Ann Arbor, Michigan, is developing environmental auditing standards that are intended to be compatible with and augment the ISO standards. Work is proceeding rapidly within ISO and NSF, with draft standards expected by the end of the year.

The proposed NSF and ISO auditing standards are being developed within the framework of overall environmental management systems standards. Neither ISO nor NSF intends to establish specific environmental standards; instead both are seeking to provide management tools that include auditing schemes and standards. The EPA 1986 policy has been a central reference document for both the ISO and NSF work. As these new documents develop, issues of auditor qualifications and explicit management commitment to audit follow-up will be of particular interest to EPA.

IV. Review of EPA Policy on Specific Environmental Auditing Issues

A. Agency Requests for Audit Reports

EPA's 1986 policy clearly states that, ". . . EPA believes routine agency requests for audit reports could inhibit auditing in the long run, decreasing both the quantity and quality of audits conducted. Therefore, as a matter of policy EPA will not routinely request environmental audit reports."

The 1986 policy also acknowledges regulated entities' need to self-evaluate environmental performance with some measure of privacy. However, audit reports may not shield monitoring, compliance, or other information that would otherwise be reportable and/or accessible to EPA even if there is no explicit requirement to generate that data. Thus, the 1986 policy does not alter regulated entities' existing or future obligations to monitor, record, or report information required under environmental statutes, regulations, or permits, or to allow EPA access to that information. Nor does the 1986 policy alter EPA's authority to request and receive any relevant information — including that contained in audit reports — under various environmental statutes or in other administrative or judicial proceedings.

EPA's authority to request an audit report, or relevant portions thereof, will be exercised on a case-by-case basis where the agency determines it is needed to accomplish a statutory mission, or where the government deems it to be material to a criminal investigation. EPA expects such requests to be limited, most likely focused on particular information needs rather than the entire report, and usually made where the information needed cannot be obtained from monitoring, reporting, or other data otherwise available to the agency. Examples would likely include situations where: audits are conducted under consent decrees or other settlement agreements; a company has placed its management practices at issue by raising them as a defense; or state of mind or intent are a relevant element of inquiry, such as during a criminal investigation. This list is illustrative rather than exhaustive, since there doubtless will be other situations, not subject to prediction, in which audit reports rather than information may be required.

B. EPA Response to Environmental Auditing

1. General Policy

The 1986 policy states that, "EPA will not promise to forgo inspections, reduce enforcement responses, or offer other such incentives in exchange for implementation of environmental auditing or other sound environmental management practices." EPA is required by law to independently assess compliance status of facilities, and cannot eliminate inspections for particular firms or classes of firms. Certain statutes (e.g., RCRA) and agency policies establish minimum facility inspection frequencies to which EPA will adhere. Environmental audits are in no way a substitute for regulatory oversight.

As explained in the 1986 policy, however, EPA will take into account a facility's efforts to audit in setting inspection priorities and in fashioning enforcement responses to violations, ". . . EPA will continue to address environmental problems on a priority basis and will consequently inspect facilities with poor environmental records and practices more frequently. Since effective environmental auditing helps management identify and promptly correct actual or potential problems, audited facilities' environmental performance should improve. Thus, while EPA inspections of self-audited facilities will continue, to the extent that compliance performance is considered in setting inspection priorities, facilities with a good compliance history may be subject to fewer inspections."

In fashioning enforcement responses to violations, EPA policy is to take into account, on a case-by-case basis, the honest and genuine efforts of regulated entities to avoid and promptly correct violations and underlying environmental problems. When regulated entities take reasonable precautions to avoid noncompliance, expeditiously correct underlying environmental problems discovered through audits or other means, and implement measures to prevent their recurrence, EPA may exercise its discretion to consider such actions as honest and genuine efforts to assure compliance. Such consideration applies particularly when a regulated entity promptly reports violations or compliance data that otherwise were not required to be recorded or reported to EPA.

These principles have been incorporated into the agency's enforcement response and civil penalty policies.

2. Audit Provisions as Remedies in Enforcement Settlements

The 1986 policy includes the following language on audit provisions as remedies in enforcement settlements:

EPA may propose environmental auditing provisions in consent decrees and in other settlement negotiations where auditing could provide a remedy for identified problems and reduce the likelihood of similar problems recurring in the future. Environmental auditing provisions are most likely to be proposed in settlement negotiations when:

- A pattern of violations can be attributed, at least in part, to the absence or poor functioning of an environmental management system; or
- The type or nature of violations indicates a likelihood that similar noncompliance problems may exist or occur elsewhere in the facility or at other facilities operated by the regulated entity.

EPA's enforcement office issued further guidance on this issue in 1986 in a document entitled *EPA Policy on the Inclusion of Environmental Auditing Provisions in Enforcement Settlements.* This guidance has been consistently applied in enforcement actions as appropriate, and has formed the basis for the inclusion of audit agreements or provisions in numerous consent decrees. Selected text from this document, also still in effect, is included here:

In recent years, agency negotiators have achieved numerous settlements that require regulated entities to audit their operations. These innovative settlements have been highly successful in enabling the agency to accomplish more effectively its primary mission, namely, to secure environmental compliance. Indeed, auditing provisions in enforcement settlements have provided several important benefits to the agency by enhancing its ability to:

- Address compliance at an entire facility or at all facilities owned or operated by a party, rather than just the violations discovered during inspections; and identify and correct violations that may have gone undetected (and uncorrected) otherwise;
- Focus the attention of a regulated party's top-level management on environmental compliance; produce corporate policies and procedures that enable a party to achieve and maintain compliance; and help a party to manage pollution control affirmatively over time instead of reacting to crises;
- Provide a quality assurance check by verifying that existing environmental management practices are in place, functioning, and adequate.

It is the policy of EPA to settle its judicial and administrative enforcement cases only where violators can assure the agency that their noncompliance will be (or has been) corrected. EPA . . . considers auditing an appropriate part of a settlement where heightened management attention could lower the potential for noncompliance to recur.

In most cases, either (or both) of the following two types of environmental audits should be considered (in enforcement settlements):

(a) *Compliance Audit:* An independent assessment of the current status of a party's compliance with applicable statutory and regulatory requirements. This approach always entails a requirement that

effective measures be taken to remedy uncovered compliance problems, and is most effective when coupled with a requirement that the root causes of noncompliance also be remedied.
(b) *Management Audit:* An independent evaluation of a party's environmental compliance policies, practices, and controls. Such evaluation may encompass the need for: (1) A formal corporate environmental compliance policy, and procedures for implementation of that policy; (2) educational and training programs for employees; (3) equipment purchase, operation, and maintenance programs; (4) environmental compliance officer programs (or other organizational structures relevant to compliance); (5) budgeting and planning systems for environmental compliance; (6) monitoring, recordkeeping, and reporting systems; (7) in-plant and community emergency plans; (8) internal communications and control systems; and (9) hazard identification and risk assessment.

Whether to seek a compliance audit, a management audit, or both will depend upon the unique circumstances of each case. A compliance audit usually will be appropriate where the violations uncovered by agency inspections raise the likelihood that environmental noncompliance exists elsewhere within a party's operations. A management audit should be sought where it appears that a major contributing factor to noncompliance is inadequate (or nonexistent) managerial attention to environmental policies, procedures, or staffing. Both types of audits should be sought where both current noncompliance and shortcomings in a party's environmental management practices need to be addressed.

C. Environmental Auditing and Criminal Enforcement Policy

Following EPA's 1986 policy, three significant developments mark the evolution and implementation of criminal enforcement policy governing the use of self-audits and the voluntary disclosure of environmental violations.

First, on July 1, 1991, the Department of Justice issued a guidance entitled *Factors in Decisions on Criminal Prosecutions for Environmental Violations in the Context of Significant Voluntary Compliance or Disclosure Efforts by the Violator.* The guidance sets the general DOJ policy on auditing: It is the policy of the Department of Justice to encourage self-auditing, self-policing, and voluntary disclosure of environmental violations by the regulated community by indicating that these activities are viewed as mitigating factors in the department's exercise of criminal enforcement discretion.

The guidance and the examples contained therein provide a framework for the determination of whether a particular case presents the type of circumstances in which lenience would be appropriate. The factors to be considered in exercising the department's prosecutorial discretion, in cases where the law and evidence are otherwise sufficient for prosecution, include: voluntary disclosure; cooperation; preventive measures and compliance programs; pervasiveness of noncompliance; internal disciplinary action; and subsequent compliance efforts.

Second, on November 11, 1993, the Final Draft Environmental Sentencing Guidelines provided for the mitigation of sentences where a court finds that the following factors for environmental compliance are satisfied: line management attention to compliance; integration of environmental policies, standards, and procedures; auditing, monitoring, reporting, and tracking systems; regulatory expertise, training, and evaluation; incentives for compliance; disciplinary procedures; and continuing evaluation and improvement.

Finally, on January 12, 1994, EPA's Director of Criminal Enforcement issued a guidance entitled *The Exercise of Investigative Discretion,* that sets forth specific factors that distinguish cases meriting criminal investigation. With respect to corporations conducting environmental audits, the guidance states:

Corporate culpability may be indicated when a company performs an environmental compliance or management audit, and then knowingly fails to promptly remedy the noncompliance and correct any harm done. On the other hand, EPA policy strongly encourages self-monitoring, self-disclosure, and self-correction. When self-auditing has been conducted (followed up by prompt remediation of the noncompliance and any resulting harm) and full, complete disclosure has occurred, the company's constructive activities should be considered as mitigating factors in EPA's exercise of investigative discretion. Therefore, a violation that is voluntarily revealed and fully and promptly remediated as part of a corporation's systematic and comprehensive self-evaluation program generally will not be a candidate for the expenditure of scarce criminal resources.

D. Audit Privilege Legislation

Four States (Colorado, Indiana, Kentucky, and Oregon) have recently enacted legislation that, with some variations, creates a "self-evaluative" privilege for audit reports. EPA has consistently opposed this approach, principally because of the risk of weakening state enforcement programs, the imposition of unnecessary transaction costs and delays in enforcement actions, and the potential increase in the number of situations requiring the expenditure of scarce agency resources, including the "overfiling" of state enforcement actions. EPA urges states that are considering a privilege-oriented approach to actively participate in the comprehensive process described in the June 20, 1994 *Federal Register* notice (59 FR 31914) before pursuing any legislative action. The agency also encourages states that have passed such legislation to present documentary justification for this approach either at the public meeting on July 27–28, 1994, or in written comments.

E. Environmental Auditing at Federal Facilities

The 1986 policy also encourages all federal agencies subject to environmental laws and regulations to institute environmental auditing, to help ensure the adequacy of internal systems to achieve, maintain, and monitor compliance. Such federal facility environmental audit programs should be structured to promptly identify environmental problems and expeditiously develop schedules for remedial action.

Where appropriate, EPA will enter into agreements with other agencies to clarify the respective roles, responsibilities, and commitments of each agency in conducting and responding to federal facility environmental audits. Also, to the extent feasible, EPA will provide technical assistance to help federal agencies design and initiate audit programs. Currently, the EPA Federal Facility Enforcement Office (FFEO) is cochairing an interagency work group to revise auditing guidelines and protocols for federal agencies. In addition, FFEO is developing the Federal Government Environmental Challenge Program required by Executive Order 12856, which calls for the establishment of a Code of Environmental Principles and a Model Installation Program for federal facilities. This program is likely to include environmental auditing components.

The 1986 policy states that, "With respect to inspections of self-audited facilities and requests for audit reports, EPA generally will respond to environmental audits by federal facilities in the same manner as it does for other regulated entities."

Federal agencies should, however, be aware that the Freedom of Information Act will govern any disclosure of audit reports or audit-generated information requested from federal agencies by the public. When federal agencies discover significant violations through an audit, EPA encourages them to voluntarily submit the related findings and corrective action plans to the appropriate EPA regional office and state agencies, even when not specifically required to do so. EPA will review the audit findings and action plans, and negotiate either a consent agreement or a Federal Facilities Compliance Agreement, pursuant to its enforcement authorities under the various environmental statutes. In any event, federal agencies are expected to report to EPA pollution abatement and prevention projects involving costs necessary to correct compliance problems discovered through the audit, in accordance with OMB Circular A-106. Upon request, and in appropriate circumstances, EPA may assist affected federal agencies through coordination of any public release of voluntarily submitted audit findings with approved action plans once agreement has been reached and/or appropriate enforcement actions have been taken.

V. Review of Relationship to State or Local Regulatory Agencies

Effective federal/state partnerships are critical to accomplishing the mutual goal of achieving and maintaining high levels of compliance with environmental laws and regulations. The greater the consistency between state and local policies and the federal response to environmental auditing, the greater the degree to which sound auditing practices might be adopted and compliance levels improved. State and local regulatory agencies, of course, have independent jurisdiction over regulated entities. EPA encourages them to adopt these or similar policies on environmental auditing, in order to advance the use of effective environmental auditing in a consistent manner.

The 1986 policy emphasizes this point further: EPA recognizes that some states have already undertaken environmental auditing initiatives that differ somewhat from this policy. Other states also may want to develop auditing policies that accommodate their particular needs or circumstances. Nothing in this policy statement is intended to preempt or preclude states from developing other approaches to environmental auditing. EPA encourages state and local authorities to consider the basic principles that guided the agency in developing this policy:

- Regulated entities must continue to report or record compliance information required under existing statutes or regulations, regardless of whether such information is generated by an environmental audit or contained in an audit report. Required information cannot be withheld merely because it is generated by an audit rather than by some other means.
- Regulatory agencies cannot make promises to forgo or limit enforcement action against a particular facility or class of facilities in exchange for the use of environmental auditing systems. However, such agencies may use their discretion to adjust enforcement actions on a case-by-case basis in response to honest and genuine efforts by regulated entities to assure environmental compliance.
- When setting inspection priorities, regulatory agencies should focus to the extent possible on compliance performance and environmental results.
- Regulatory agencies must continue to meet minimum program requirements (e.g., minimum inspection requirements, etc.).
- Regulatory agencies should not attempt to prescribe the precise form and structure of regulated entities' environmental management or auditing programs.

VI. Conclusion

All of the policies referenced in this notice remain in effect. The agency intends, however, to reexamine these policies comprehensively and remains open to suggestions for changes and improvements regarding all aspects of existing auditing policy. The information presented here is intended for the convenience of interested parties, in preparation for the July 27–28, 1994, public meeting. The agency hopes that this information will clarify EPA's current activities and policies related to environmental auditing.

APPENDIX A

The Office of Compliance will respond to written requests for copies of the documents referenced in this notice. Send all requests to: U.S. EPA, Office of Compliance, Attn: Ira R. Feldman, Special Counsel, 401 M Street, NW (5503), Washington, DC 20460.

Steven A. Herman, Assistant Administrator, Office of Enforcement and Compliance Assurance

Federal Register / Vol. 60, No. 246 / Friday, December 22, 1995 / Notices

INCENTIVES FOR SELF-POLICING: DISCOVERY, DISCLOSURE, CORRECTION, AND PREVENTION OF VIOLATIONS

Agency: Environmental Protection Agency (EPA).
Action: Final Policy Statement.

Summary: The Environmental Protection Agency (EPA) today issues its final policy to enhance protection of human health and the environment by encouraging regulated entities to voluntarily discover, and disclose and correct violations of environmental requirements. Incentives include eliminating or substantially reducing the gravity component of civil penalties and not recommending cases for criminal prosecution where specified conditions are met, for those who voluntarily self-disclose and promptly correct violations. The policy also restates EPA's long-standing practice of not requesting voluntary audit reports to trigger enforcement investigations. This policy was developed in close consultation with the U.S. Department of Justice, states, public interest groups, and the regulated community and will be applied uniformly by the agency's enforcement programs.

Dates: This policy is effective January 22, 1996.

For further information contact: Additional documentation relating to the development of this policy is contained in the environmental auditing public docket. Documents from the docket may be obtained by calling (202) 260-7548, requesting an index to docket #C-94-01, and faxing document requests to (202) 260-4400. Hours of operation are 8 A.M. to 5:30 P.M., Monday through Friday, except legal holidays. Additional contacts are Robert Fentress or Brian Riedel at (202) 564-4187.

SUPPLEMENTARY INFORMATION:

I. Explanation of Policy

A. Introduction

The Environmental Protection Agency today issues its final policy to enhance protection of human health and the environment by encouraging regulated entities to discover voluntarily, disclose, correct, and prevent violations of federal environmental law. Effective 30 days from today, where violations are found through voluntary environmental audits or efforts that reflect a regulated entity's due diligence, and are promptly disclosed and expeditiously corrected, EPA will not seek gravity-based (i.e., noneconomic benefit) penalties and will generally not recommend criminal prosecution against the regulated entity. EPA will reduce gravity-based penalties by 75% for violations that are voluntarily discovered, and are promptly disclosed and corrected, even

APPENDIX A 171

if not found through a formal audit or due diligence. Finally, the policy restates EPA's long-held policy and practice to refrain from routine requests for environmental audit reports.

The policy includes important safeguards to deter irresponsible behavior and protect the public and environment. For example, in addition to prompt disclosure and expeditious correction, the policy requires companies to act to prevent recurrence of the violation and to remedy any environmental harm that may have occurred. Repeated violations or those that result in actual harm or may present imminent and substantial endangerment are not eligible for relief under this policy, and companies will not be allowed to gain an economic advantage over their competitors by delaying their investment in compliance. Corporations remain criminally liable for violations that result from conscious disregard of their obligations under the law, and individuals are liable for criminal misconduct.

The issuance of this policy concludes EPA's 18-month public evaluation of the optimum way to encourage voluntary self-policing while preserving fair and effective enforcement. The incentives, conditions, and exceptions announced today reflect thoughtful suggestions from the Department of Justice, state attorneys general and local prosecutors, state environmental agencies, the regulated community, and public interest organizations. EPA believes that it has found a balanced and responsible approach and will conduct a study within 3 years to determine the effectiveness of this policy.

B. Public Process

One of the Environmental Protection Agency's most important responsibilities is ensuring compliance with federal laws that protect public health and safeguard the environment. Effective deterrence requires inspecting, bringing penalty actions, and securing compliance and remediation of harm. But EPA realizes that achieving compliance also requires the cooperation of thousands of businesses and other regulated entities subject to these requirements. Accordingly, in May of 1994, the administrator asked the Office of Enforcement and Compliance Assurance (OECA) to determine whether additional incentives were needed to encourage voluntary disclosure and correction of violations uncovered during environmental audits.

EPA began its evaluation with a 2-day public meeting in July of 1994, in Washington, D.C., followed by a 2-day meeting in San Francisco on January 19, 1995, with stakeholders from industry, trade groups, state environmental commissioners and attorneys general, district attorneys, public interest organizations and professional environmental auditors. The agency also established and maintained a public docket of testimony presented at these meetings and all comments and correspondence submitted to EPA by outside parties on this issue.

In addition to considering opinions and information from stakeholders, the agency examined other federal and state policies related to self-policing, self-disclosure, and correction. The agency also considered relevant surveys on auditing practices in the private sector. EPA completed the first stage of this effort with the announcement of an interim policy on April 3 of this year, which defined conditions under which EPA would reduce civil penalties and not recommend criminal prosecution for companies that audited, disclosed, and corrected violations.

Interested parties were asked to submit comments on the interim policy by June 30 of this year (60 FR 16875), and EPA received over 300 responses from a wide variety of private and public organizations. (Comments on the interim audit policy are contained in the Auditing Policy Docket, hereinafter, "Docket.") Further, the American Bar Association SONREEL Subcommittee hosted 5 days of dialogue with representatives from the regulated industry, states, and public interest organizations in June and September of this year, which identified options for strengthening the interim policy. The changes to the interim policy announced today reflect insight gained through comments submitted to EPA, the ABA dialogue, and the agency's practical experience implementing the interim policy.

C. Purpose

This policy is designed to encourage greater compliance with laws and regulations that protect human health and the environment. It promotes a higher standard of self-policing by waiving gravity-based penalties for violations that are promptly disclosed and corrected, and which were discovered through voluntary audits or compliance management systems that demonstrate due diligence. To further promote compliance, the policy reduces gravity-based penalties by 75% for any violation voluntarily discovered and promptly disclosed and corrected, even if not found through an audit or compliance management system.

EPA's enforcement program provides a strong incentive for responsible behavior by imposing stiff sanctions for noncompliance. Enforcement has contributed to the dramatic expansion of environmental auditing measured in numerous recent surveys. For example, more than 90% of the corporate respondents to a 1995 Price Waterhouse survey who conduct audits said that one of the reasons they did so was to find and correct violations before they were found by government inspectors. (A copy of the Price Waterhouse survey is contained in the Docket as document VIII-A-76.)

At the same time, because government resources are limited, maximum compliance cannot be achieved without active efforts by the regulated community to police themselves. More than half of the respondents to the same 1995 Price Waterhouse survey said that they would expand environmental auditing in exchange for reduced penalties for violations discovered and corrected. While many companies already audit or have compliance management programs, EPA believes that the incentives offered in this policy will improve the frequency and quality of these self-monitoring efforts.

D. Incentives for Self-Policing

Section C of EPA's policy identifies the major incentives that EPA will provide to encourage self-policing, self-disclosure, and prompt self-correction. These include not seeking gravity-based civil penalties or reducing them by 75%, declining to recommend criminal prosecution for regulated entities that self-police, and refraining from routine requests for audits. (As noted in Section C of the policy, EPA has refrained from making routine requests for audit reports since issuance of its 1986 policy on environmental auditing.)

1. Eliminating Gravity-Based Penalties

Under Section C(1) of the policy, EPA will not seek gravity-based penalties for violations found through auditing that are promptly disclosed and corrected. Gravity-based penalties will also be waived for violations found through any documented procedure for self-policing, where the company can show that it has a compliance management program that meets the criteria for due diligence in Section B of the policy.

Gravity-based penalties (defined in Section B of the policy) generally reflect the seriousness of the violator's behavior. EPA has elected to waive such penalties for violations discovered through due diligence or environmental audits, recognizing that these voluntary efforts play a critical role in protecting human health and the environment by identifying, correcting, and ultimately preventing violations. All of the conditions set forth in Section D, which include prompt disclosure and expeditious correction, must be satisfied for gravity-based penalties to be waived.

As in the interim policy, EPA reserves the right to collect any economic benefit that may have been realized as a result of noncompliance, even where companies meet all other conditions of the policy. Economic benefit may be waived, however, where the agency determines that it is insignificant.

After considering public comment, EPA has decided to retain the discretion to recover economic benefit for two reasons. First, it provides an incentive to comply on time. Taxpayers expect to pay interest or a penalty fee if their tax payments are late; the same principle should apply to corporations that have delayed their investment in compliance. Second, it is fair because it protects responsible companies from being undercut by their noncomplying competitors, thereby preserving a level playing field. The concept of recovering economic benefit was supported in public comments by many stakeholders, including industry representatives (see, e.g., Docket, U-F-39, 11-F-28, and B-F-18).

2. 75% Reduction of Gravity

The policy appropriately limits the complete waiver of gravity-based civil penalties to companies that meet the higher standard of environmental auditing or systematic compliance management. However, to provide additional encouragement for the kind of self-policing that benefits the public, gravity-based penalties will be reduced by 75% for a violation that is voluntarily discovered, promptly disclosed, and expeditiously corrected, even if it was not found through an environmental audit and the company cannot document due diligence. EPA expects that this will encourage companies to come forward and work with the agency to resolve environmental problems and begin to develop an effective compliance management program.

Gravity-based penalties will be reduced 75% only where the company meets all conditions in Sections D(2) through D(9). EPA has eliminated language from the interim policy indicating that penalties may be reduced "up to" 75% where "most" conditions are met, because the agency believes that all of the conditions in D(2) through D(9) are reasonable and essential to achieving compliance. This change also responds to requests for greater clarity and predictability.

3. No Recommendations for Criminal Prosecution

EPA has never recommended criminal prosecution of a regulated entity based on voluntary disclosure of violations discovered through audits and disclosed to the government before an investigation was already under way. Thus, EPA will not recommend criminal prosecution for a regulated entity that uncovers violations through environmental audits or due diligence, promptly discloses and expeditiously corrects those violations, and meets all other conditions of Section D of the policy.

This policy is limited to good actors and therefore has important limitations. It will not apply, for example, where corporate officials are consciously involved in or willfully blind to violations, or conceal or condone noncompliance. Since the regulated entity must satisfy all of the conditions of Section D of the policy, violations that caused serious harm or that may pose imminent and substantial endangerment to human health or the environment are not covered by this policy. Finally, EPA reserves the right to recommend prosecution for the criminal conduct of any culpable individual.

Even where all of the conditions of this policy are not met, however, it is important to remember that EPA may decline to recommend prosecution of a company or individual for many other reasons under other agency enforcement policies. For example, the agency may decline to recommend prosecution where there is no significant harm or culpability and the individual or corporate defendant has cooperated fully.

Where a company has met the conditions for avoiding a recommendation for criminal prosecution under this policy, it will not face any civil liability for gravity-based penalties. That is because the same conditions for discovery, disclosure, and correction apply in both cases. This represents a clarification of the interim policy, not a substantive change.

4. No Routine Requests for Audits

EPA is reaffirming its policy, in effect since 1986, to refrain from routine requests for audits. Eighteen months of public testimony and debate have produced no evidence that the agency has deviated, or should deviate, from this policy.

If the agency has independent evidence of a violation, it may seek information needed to establish the extent and nature of the problem and the degree of culpability. In general, however, an audit that results in prompt correction clearly will reduce liability, not expand it. Furthermore, a review of the criminal docket did not reveal a single criminal prosecution for violations discovered as a result of an audit self-disclosed to the government.

E. Conditions

Section D describes the nine conditions that a regulated entity must meet in order for the agency not to seek (or to reduce) gravity-based penalties under the policy. As explained in the Summary above, regulated entities that meet all nine conditions will not face gravity-based civil penalties, and will generally not have to fear criminal prosecution. Where the regulated entity meets all of the conditions except the first (D[1]), EPA will reduce gravity-based penalties by 75%.

1. Discovery of the Violation Through an Environmental Audit or Due Diligence

Under Section D(1), the violation must have been discovered through either (a) an environmental audit that is systematic, objective, and periodic as defined in the 1986 audit policy, or (b) a documented, systematic procedure or practice that reflects the regulated entity's due diligence in preventing, detecting, and correcting violations. The interim policy provided full credit for any violation found through "voluntary self-evaluation," even if the evaluation did not constitute an audit. In order to receive full credit under the final policy, any self-evaluation that is not an audit must be part of a "due diligence" program. Both "environmental audit" and "due diligence" are defined in Section B of the policy.

Where the violation is discovered through a "systematic procedure or practice" that is not an audit, the regulated entity will be asked to document how its program reflects the criteria for due diligence as defined in Section B of the policy. These criteria, which are adapted from existing codes of practice such as the 1991 Criminal Sentencing Guidelines, were fully discussed during the ABA dialogue. The criteria are flexible enough to accommodate different types and sizes of businesses. The agency recognizes that a variety of compliance management programs may develop under the due diligence criteria, and will use its review under this policy to determine whether basic criteria have been met.

Compliance management programs that train and motivate production staff to prevent, detect, and correct violations on a daily basis are a valuable complement to periodic auditing. The policy is responsive to recommendations received during public comment and from the ABA dialogue to give compliance management efforts that meet the criteria for due diligence the same penalty reduction offered for environmental audits (see, e.g., II-F-39, II-E-18, and II-G-18 in the Docket).

EPA may require as a condition of penalty mitigation that a description of the regulated entity's due diligence efforts be made publicly available. The agency added this provision in response to suggestions from environmental groups, and believes that the availability of such information will allow the public to judge the adequacy of compliance management systems, lead to enhanced compliance, and foster greater public trust in the integrity of compliance management systems.

2. Voluntary Discovery and Prompt Disclosure

Under Section D(2) of the final policy, the violation must have been identified voluntarily, and not through a monitoring, sampling, or auditing procedure that is required by statute, regulation, permit, judicial or administrative order, or consent agreement. Section D(4) requires that disclosure of the violation be prompt and in writing. To avoid confusion and respond to state requests for greater clarity, disclosures under this policy should be made to EPA. The agency will work closely with states in implementing the policy.

The requirement that discovery of the violation be voluntary is consistent with proposed federal and state bills that would reward those discoveries that the regulated entity can legitimately attribute to its own voluntary efforts.

The policy gives three specific examples of discovery that would not be voluntary, and therefore would not be eligible for penalty mitigation: emissions violations detected through a required continuous emissions monitor, violations of NPDES discharge limits found through prescribed monitoring, and violations discovered through a compliance audit required to be performed by the terms of a consent order or settlement agreement.

The final policy generally applies to any violation that is voluntarily discovered, regardless of whether the violation is required to be reported. This definition responds to comments pointing out that reporting requirements are extensive, and that excluding them from the policy's scope would severely limit the incentive for self-policing (see, e.g., II-C-48 in the Docket).

The agency wishes to emphasize that the integrity of federal environmental law depends upon timely and accurate reporting. The public relies on timely and accurate reports from the regulated community not only to measure compliance, but to evaluate health or environmental risk and gauge progress in reducing pollutant loadings. EPA expects the policy to encourage the kind of vigorous self-policing that will serve these objectives, and not to provide an excuse for delayed reporting. Where violations of reporting requirements are voluntarily discovered, they must be promptly reported (as discussed below). Where a failure to report results in imminent and substantial

APPENDIX A

endangerment or serious harm, that violation is not covered under this policy (see Condition D[8]). The policy also requires the regulated entity to prevent recurrence of the violation to ensure that noncompliance with reporting requirements is not repeated. EPA will closely scrutinize the effect of the policy in furthering the public interest in timely and accurate reports from the regulated community.

Under Section D(4), disclosure of the violation should be made within 10 days of its discovery, and in writing to EPA. Where a statute or regulation requires reporting be made in less than 10 days, disclosure should be made within the time limit established by law. Where reporting within 10 days is not practical because the violation is complex and compliance cannot be determined within that period, the agency may accept later disclosures if the circumstances do not present a serious threat and the regulated entity meets its burden of showing that the additional time was needed to determine compliance status.

This condition recognizes that it is critical for EPA to get timely reporting of violations in order that it might have clear notice of the violations and the opportunity to respond if necessary, as well as an accurate picture of a given facility's compliance record. Prompt disclosure is also evidence of the regulated entity's good faith in wanting to achieve or return to compliance as soon as possible.

In the final policy, the agency has added the words, "or may have occurred," to the sentence, "The regulated entity fully discloses that a specific violation has occurred, or may have occurred . . . ." This change, which was made in response to comments received, clarifies that where an entity has some doubt about the existence of a violation, the recommended course is for it to disclose and allow the regulatory authorities to make a definitive determination.

In general, the Freedom of Information Act will govern the agency's release of disclosures made pursuant to this policy. EPA will, independently of FOIA, make publicly available any compliance agreements reached under the policy (see Section H of the policy), as well as descriptions of due diligence programs submitted under Section D. 1 of the Policy. Any material claimed to be Confidential Business Information will be treated in accordance with EPA regulations at 40 CFR Part 2.

3. Discovery and Disclosure Independent of Government or Third-Party Plaintiff

Under Section D(3), in order to be . . . "voluntary", the violation must be identified and disclosed by the regulated entity prior to: the commencement of a federal, state, or local agency inspection, investigation, or information request; notice of a citizen suit; legal complaint by a third party; the reporting of the violation to EPA by a "whistle-blower" employee; and imminent discovery of the violation by a regulator agency.

This condition means that regulated entities must have taken the initiative to find violations and promptly report them, rather than reacting to knowledge of a pending enforcement action or third-party complaint. This concept was reflected in the interim policy and in federal and state penalty immunity laws and did not prove controversial in the public comment process.

4. Correction and Remediation

Section D(5) ensures that, in order to receive the penalty mitigation benefits available under the policy, the regulated entity not only voluntarily discovers and promptly discloses a violation, but expeditiously corrects it, remedies any harm caused by that violation (including responding to any spill and carrying out any removal or remedial action required by law), and expeditiously certifies in writing to appropriate state, local, and EPA authorities that violations have been corrected. It also enables EPA to ensure that the regulated entity will be publicly accountable for its commitments through binding written agreements, orders, or consent decrees where necessary.

The final policy requires the violation to be corrected within 60 days, or that the regulated entity provide written notice where violations may take longer to correct. EPA recognizes that some violations can and should be corrected immediately, while others (e.g., where capital expenditures are involved) may take longer than 60 days to correct. In all cases, the regulated entity will be expected to do its utmost to achieve or return to compliance as expeditiously as possible.

Where correction of the violation depends upon issuance of a permit that has been applied for but not issued by federal or state authorities, the agency will, where appropriate, make reasonable efforts to secure timely review of the permit.

5. Prevent Recurrence

Under Section D(6), the regulated entity must agree to take steps to prevent a recurrence of the violation, including but not limited to improvements to its environmental auditing or due diligence efforts. The final policy makes clear that the preventive steps may include improvements to a regulated entity's environmental auditing or due diligence efforts to prevent recurrence of the violation.

In the interim policy, the agency required that the entity implement appropriate measures to prevent a recurrence of the violation, a requirement that operates prospectively. However, a separate condition in the interim policy also required that the violation not indicate "a failure to take appropriate steps to avoid repeat or recurring violations" — a requirement that operates retrospectively. In the interest of both clarity and fairness, the agency has decided for purposes of this condition to keep the focus prospective and thus to require only that steps be taken to prevent recurrence of the violation after it has been disclosed.

6. No Repeat Violations

In response to requests from commenters (see, e.g., 11-F-39 and II-C-18 in the Docket), EPA has established "bright lines" to determine when previous violations will bar a regulated entity from obtaining relief under this policy. Those will help protect the public and responsible companies by ensuring that penalties are not waived for repeat offenders. Under condition D(7), the same or closely-related violation must not have occurred previously within the past 3 years at the same facility, or be part of a pattern of violations on the regulated entity's part over the past 5 years. This provides companies with a continuing incentive to prevent violations, without being unfair to regulated entities responsible for managing hundreds of facilities. It would be unreasonable to provide unlimited amnesty for repeated violations of the same requirement.

The term "violation" includes any violation subject to a federal or state civil, judicial, or administrative order, consent agreement, conviction, or plea agreement. Recognizing that minor violations are sometimes settled without a formal action in court, the term also covers any act or omission for which the regulated entity has received a penalty reduction in the past. Together, these conditions identify situations in which the regulated community has had clear notice of its noncompliance and an opportunity to correct.

7. Other Violations Excluded

Section D(8) makes clear that penalty reductions are not available under this policy for violations that resulted in serious actual harm or that may have presented an imminent and substantial endangerment to public health or the environment. Such events indicate a serious failure (or absence) of a self-policing program, which should be designed to prevent such risks, and it would seriously undermine deterrence to waive penalties for such violations. These exceptions are responsive to suggestions from public interest organizations, as well as other commenters (see, e.g., II-F-39 and II-G-18 in the Docket).

The final policy also excludes penalty reductions for violations of the specific terms of any order, consent agreement, or plea agreement (see II-E-60 in the Docket). Once a consent agreement has been negotiated, there is little incentive to comply if there are no sanctions for violating its specific requirements. The exclusion in this section applies to violations of the terms of any response, removal, or remedial action covered by a written agreement.

8. Cooperation

Under Section D(9), the regulated entity must cooperate as required by EPA and provide information necessary to determine the applicability of the policy. This condition is largely unchanged from the interim policy. In the final policy, however, the agency has added that "cooperation" includes assistance in determining the facts of any related violations suggested by the disclosure, as well as of the disclosed violation itself. This was added to allow the agency to obtain information about any violations indicated by the disclosure, even where the violation is not initially identified by the regulated entity.

F. Opposition to Privilege

The agency remains firmly opposed to the establishment of a statutory evidentiary privilege for environmental audits for the following reasons:

1. Privilege, by definition, invites secrecy, instead of the openness needed to build public trust in industry's ability to self-police. American law reflects the high value that the public places on fair access to the facts. The Supreme Court, for example, has said of privileges that, "[w]hatever their origins, these exceptions to the demand for every man's evidence are not lightly created nor expansively construed, for they are in derogation of the search for truth." *United States v. Nixon,* 418 U.S. 683 (1974). Federal courts have unanimously refused to recognize a privilege for environmental audits in the context of government investigations. See, e.g., *United States v. Dexter,* 132 F.R.D. 8, 9-10 (D. Conn. 1990) (application of a privilege "would effectively impede [EPA's] ability to enforce the Clean Water Act" and would be contrary to stated public policy).
2. Eighteen months have failed to produce any evidence that a privilege is needed. Public testimony on the interim policy confirmed that EPA rarely uses' audit reports as evidence. Furthermore, surveys demonstrate that environmental auditing has expanded rapidly over the past decade without the stimulus of a privilege. Most recently, the 1995 Price Waterhouse Survey found that those few large or mid-sized companies that do not audit generally do not perceive any need to; concern about confidentiality ranked as one of the least important factors in their decisions.
3. A privilege would invite defendants to claim as "audit" material almost any evidence the government needed to establish a violation or determine who was responsible. For example, most audit privilege bills under consideration in federal and state legislatures would arguably protect factual information, such as health studies or contaminated sediment data, and not just the conclusions of the auditors. While the government might have access to required monitoring data under the law, as some industry commenters have suggested, a privilege of that nature would cloak underlying facts needed to determine whether such data were accurate.
4. An audit privilege would breed litigation, as both parties struggled to determine what material fell within its scope. The problem is compounded by the lack of any clear national standard for audits. The "in camera" (i.e., nonpublic) proceedings used to resolve these

APPENDIX A 181

disputes under some statutory schemes would result in a series of time-consuming, expensive mini-trials.
5. The agency's policy eliminates the need for any privilege as against the government, by reducing civil penalties and criminal liability for those companies that audit, disclose, and correct violations. The 1995 Price Waterhouse survey indicated that companies would expand their auditing programs in exchange for the kind of incentives that EPA provides in its policy.
6. Finally, audit privileges are strongly opposed by the law enforcement community, including the National District Attorneys Association, as well as by public interest groups. (See, e.g., Docket, II-C-21, II-C-28, II-C-52, IV-C-10, II-C-25, II-C-33, II-C-52, II-C-48, and II-G-13 through II-G-24.)

G. Effect on States

The final policy reflects EPA's desire to develop fair and effective incentives for self-policing that will have practical value to states that share responsibility for enforcing federal environmental laws. To that end, the agency has consulted closely with state officials in developing this policy, through a series of special meetings and conference calls, in addition to the extensive opportunity for public comment. As a result, EPA believes its final policy is grounded in commonsense principles that should prove useful in the development of state programs and policies.

As always, states are encouraged to experiment with different approaches that do not jeopardize the fundamental national interest in assuring that violations of federal law do not threaten the public health or the environment, or make it profitable not to comply. The agency remains opposed to state legislation that does not include these basic protections, and reserves its right to bring independent action against regulated entities for violations of federal law that threaten human health or the environment, reflect criminal conduct or repeated noncompliance, or allow one company to make a substantial profit at the expense of its law-abiding competitors. Where a state has obtained appropriate sanctions needed to deter such misconduct, there is no need for EPA action.

H. Scope of Policy

EPA has developed this document as a policy to guide settlement actions. EPA employees will be expected to follow this policy, and the agency will take steps to assure national consistency in application. For example, the agency will make public any compliance agreements reached under this policy, in order to provide the regulated community with fair notice of decisions and greater accountability to affected communities. Many in the regulated community recommended that the agency convert the policy into a regulation because they felt it might ensure greater consistency and predictability. While EPA is taking steps to ensure consistency and predictability and believes that it will be successful, the agency will consider this issue and will provide notice if it determines that a rulemaking is appropriate.

II. Statement of Policy: Incentives for Self-Policing

A. Purpose

This policy is designed to enhance protection of human health and the environment by encouraging regulated entities to voluntarily discover, disclose, correct, and prevent violations of federal environmental requirements.

B. Definitions

For purposes of this policy, the following definitions apply:

"Environmental Audit" has the definition given to it in EPA's 1986 audit policy on environmental auditing, i.e., "a systematic, documented, periodic and objective review by regulated entities of facility operations and practices related to meeting environmental requirements."

"Due Diligence" encompasses the regulated entity's systematic efforts, appropriate to the size and nature of its business, to prevent, detect, and correct violations through all of the following:

(a) Compliance policies, standards, and procedures that identify how employees and agents are to meet the requirements of laws, regulations, permits, and other sources of authority for environmental requirements;
(b) Assignment of overall responsibility for overseeing compliance with policies, standards, and procedures, and assignment of specific responsibility for assuring compliance at each facility or operation;
(c) Mechanisms for systematically assuring that compliance policies, standards, and procedures are being carried out, including monitoring and auditing systems reasonably designed to detect and correct violations, periodic evaluation of the overall performance of the compliance management system, and a means for employees or agents to report violations of environmental requirements without fear of retaliation;
(d) Efforts to communicate effectively the regulated entity's standards and procedures to all employees and other agents;
(e) Appropriate incentives to managers and employees to perform in accordance with the compliance policies, standards, and procedures, including consistent enforcement through appropriate disciplinary mechanisms; and

APPENDIX A

(f) Procedures for the prompt and appropriate correction of any violations, and any necessary modifications to the regulated entity's program to prevent future violations.

"Environmental audit report" means the analysis, conclusions, and recommendations resulting from an environmental audit, but does not include data obtained in, or testimonial evidence concerning, the environmental audit.

"Gravity-based penalties" are that portion of a penalty over and above the economic benefit., i.e., the punitive portion of the penalty, rather than that portion representing a defendant's economic gain from noncompliance. (For further discussion of this concept, see "A Framework for Statute-Specific Approaches to Penalty Assessments," #GM-22,1980, U.S. EPA General Enforcement Policy Compendium.)

"Regulated entity" means any entity, including a federal, state, or municipal agency or facility, regulated under federal environmental laws.

C. Incentives for Self-Policing

1. No Gravity-Based Penalties

Where the regulated entity establishes that it satisfies all of the conditions of Section D of the policy, EPA will not seek gravity-based penalties for violations of federal environmental requirements.

2. Reduction of Gravity-Based Penalties by 75%

EPA will reduce gravity-based penalties for violations of federal environmental requirements by 75% so long as the regulated entity satisfies all of the conditions of Section D(2) through D(9) below.

3. No Criminal Recommendations

(a) EPA will not recommend to the Department of Justice or other prosecuting authority that criminal charges be brought against a regulated entity where EPA determines that all of the conditions in Section D are satisfied, so long as the violation does not demonstrate or involve: (i) a prevalent management philosophy or practice that concealed or condoned environmental violations; or (ii) high-level corporate officials' or managers' conscious involvement in, or willful blindness to, the violations.

(b) Whether or not EPA refers the regulated entity for criminal prosecution under this section, the agency reserves the right to recommend prosecution for the criminal acts of individual managers or employees under existing policies guiding the exercise of enforcement discretion.

4. No Routine Request for Audits

EPA will not request or use an environmental audit report to initiate a civil or criminal investigation of the entity. For example, EPA will not request an environmental audit report in routine inspections. If the agency has independent reason to believe that a violation has occurred, however, EPA may seek any information relevant to identifying violations or determining liability or extent of harm.

D. Conditions

1. Systematic Discovery

The violation was discovered through: (a) an environmental audit; or (b) an objective, documented, systematic procedure or practice reflecting the regulated entity's due diligence in preventing, detecting, and correcting violations. The regulated entity must provide accurate and complete documentation to the agency as to how it exercises due diligence to prevent, detect, and correct violations according to the criteria for due diligence outlined in Section B. EPA may require as a condition of penalty mitigation that a description of the regulated entity's due diligence efforts be made publicly available.

2. Voluntary Discovery

The violation was identified voluntarily, and not through a legally mandated monitoring or sampling requirement prescribed by statute, regulation, permit, judicial or administrative order, or consent agreement. For example, the policy does not apply to:

(a) emissions violations detected through a continuous emissions monitor (or alternative monitor established in a permit) where any such monitoring is required;

(b) violations of National Pollutant Discharge Elimination System (NPDES) discharge limits detected through required sampling or monitoring;

(c) violations discovered through a compliance audit required to be performed by the terms of a consent order or settlement agreement.

3. Prompt Disclosure

The regulated entity fully discloses a specific violation within 10 days (or such shorter period provided by law) after it has discovered that the violation has occurred, or may have occurred, in writing to EPA.

4. Discovery and Disclosure Independent of Government or Third-Party Plaintiff

The violation must also be identified and disclosed by the regulated entity prior to:

(a) the commencement of a federal, state, or local agency inspection or investigation, or the issuance by such agency of an information request to the regulated entity;

(b) notice of a citizen suit;

(c) the filing of a complaint by a third party;

(d) the reporting of the violation to EPA (or other government agency) by a "whistle-blower" employee, rather than by one authorized to speak on behalf of the regulated entity; or

(e) imminent discovery of the violation by a regulatory agency.

5. Correction and Remediation

The regulated entity corrects the violation within 60 days, certifies in writing that violations have been corrected, and takes appropriate measures as determined by EPA to remedy any environmental or human harm due to the violation. If more than 60 days will be needed to correct the violations, the regulated entity must so notify EPA in writing before the 60-day period has passed. Where appropriate, EPA may require that to satisfy conditions 5 and 6 a regulated entity enter into a publicly available written agreement, administrative consent order, or judicial consent decree, particularly where compliance or remedial measures are complex or a lengthy schedule for attaining and maintaining compliance or remediating harm is required.

6. Prevent Recurrence

The regulated entity agrees in writing to take steps to prevent a recurrence of the violation, which may include improvements to its environmental auditing or due diligence efforts.

7. No Repeat Violations

The specific violation (or closely related violation) has not occurred previously within the past 3 years at the same facility, or is not part of a pattern of federal, state, or local violations by the facility's parent organization (if any), which have occurred within the past 5 years. For the purposes of this section, a violation is:

(a) any violation of federal, state, or local environmental law identified in a judicial or administrative order, consent agreement or order, complaint, or notice of violation, conviction, or plea agreement; or

(b) any act or omission for which the regulated entity has previously received penalty mitigation from EPA or a state or local agency.

8. Other Violations Excluded

The violation is not one which (i) resulted in serious actual harm, or may have presented an imminent and substantial endangerment, to human health or the environment, or (ii) violates the specific terms of any judicial or administrative order, or consent agreement.

9. Cooperation

The regulated entity cooperates as requested by EPA and provides such information as is necessary and requested by EPA to determine applicability of this policy. Cooperation includes, at a minimum, providing all requested documents and access to employees and assistance in investigating the violation, any noncompliance problems related to the disclosure, and any environmental consequences related to the violations.

E. Economic Benefit

EPA will retain its full discretion to recover any economic benefit gained as a result of noncompliance to preserve a "level playing field" in which violators do not gain a competitive advantage over regulated entities that do comply. EPA may forgive the entire penalty for violations that meet conditions 1 through 9 in section D and, in the agency's opinion, do not merit any penalty due to the insignificant amount of any economic benefit.

F. Effect on State Law, Regulation, or Policy

EPA will work closely with states to encourage their adoption of policies that reflect the incentives and conditions outlined in this policy. EPA remains firmly opposed to statutory environmental audit privileges that shield evidence of environmental violations and undermine the public's right to know, as well as to blanket immunities for violations that reflect criminal conduct, present serious threats or actual harm to health and the environment, allow noncomplying companies to gain an economic advantage over their competitors, or reflect a repeated failure to comply with federal law. EPA will work with states to address any provisions of state audit privilege or immunity laws that are inconsistent with this policy, and which may prevent a timely and appropriate response to significant environmental violations. The agency reserves its right to take necessary actions to protect public health or the environment by enforcing against any violations of federal law.

G. Applicability

1. This policy applies to the assessment of penalties for any violations under all of the federal environmental statutes that EPA administers, and supersedes any inconsistent provisions in media-specific penalty or enforcement policies and EPA's 1986 Environmental Auditing Policy Statement.
2. To the extent that existing EPA enforcement policies are not inconsistent, they will continue to apply in conjunction with this policy. However, a regulated entity that has received penalty mitigation for satisfying specific conditions under this policy may not receive additional penalty mitigation for satisfying the same or similar conditions under other policies for the same violations, nor will this policy apply to violations that have received penalty mitigation under other policies.

APPENDIX A

3. This policy sets forth factors for consideration that will guide the agency in the exercise of its prosecutorial discretion. It states the agency's views as to the proper allocation of its enforcement resources. The policy is not final agency action, and is intended as guidance. It does not create any rights, duties, obligations, or defenses, implied or otherwise, in any third parties.
4. This policy should be used whenever applicable in settlement negotiations for both administrative and civil judicial enforcement actions. It is not intended for use in pleading, at hearing, or at trial. The policy may be applied at EPA's discretion to the settlement of administrative and judicial enforcement actions instituted prior to, but not yet resolved, as of the effective date of this policy.

H. Public Accountability

1. Within 3 years of the effective date of this policy, EPA will complete a study of the effectiveness of the policy in encouraging:
   (a) changes in compliance behavior within the regulated community, including improved compliance rates;
   (b) prompt disclosure and correction of violations, including timely and accurate compliance with reporting requirements;
   (c) corporate compliance programs that are successful in preventing violations, improving environmental performance, and promoting public disclosure;
   (d) consistency among state programs that provide incentives for voluntary compliance.
   EPA will make the study available to the public.
2. EPA will make publicly available the terms and conditions of any compliance agreement reached under this policy, including the nature of the violation, the remedy, and the schedule for returning to compliance.

I. Effective Date

This policy is effective January 22, 1996.

Dated: December 18, 1995.
Steven A. Herman, Assistant Administrator for Enforcement and Compliance Assurance

Appendix B
INTERNATIONAL STANDARDS ORGANIZATION DRAFT STANDARDS

ISO/CD 14010.2 "Guidelines for Environmental Auditing —
General Principles of Environmental Auditing" 191

ISO/CD 14011/1.2 "Guidelines for Environmental Auditing —
Audit Procedures — Part 1: Auditing of Environmental
Management Systems" 197

ISO/CD 14012.2 "Guidelines for Environmental Auditing —
Qualification Criteria for Environmental Auditors" 206

2nd Committee Draft 1 February 1995

ISO/CD 14010.2 "GUIDELINES FOR ENVIRONMENTAL AUDITING — GENERAL PRINCIPLES OF ENVIRONMENTAL AUDITING"

0 Introduction

Environmental auditing has established itself as a valuable instrument to verify and help improve environmental performance.

This International Standard is intended to guide organizations, auditors, and their clients on the general principles common to the execution of environmental audits. It provides definitions of environmental audit and related terms, and the general principles of environmental auditing:

This International Standard is one in a series of standards in the field of environmental auditing:

ISO 14010	Guidelines for environmental auditing — General principles of environmental auditing
ISO 14011/1	Guidelines for environmental auditing — Audit procedures — part 1: Auditing of environmental management systems
ISO 14012	Guidelines for environmental auditing — Qualification criteria for environmental auditors

More standards in this series may be prepared in future.

1 Scope

This International Standard provides the general principles of environmental auditing that are applicable to all types of environmental audits. Any activity defined as an environmental audit in accordance with this International Standard should satisfy the recommendations given in this International Standard.

2 Normative References

The following standards contain provisions that, through reference in this text, constitute provisions of this International Standard. At the time of publication, the editions indicated were valid. All standards are subject to revision, and parties to agreements based on International Standard are encouraged to investigate the possibility of applying the most recent editions of the standards indicated below. Members of IEC and ISO maintain registers of currently valid International Standards.

ISO 14001:199X, Environmental Management Systems — Specification with guidance for use

ISO 14011/1:199X, Guidelines for environmental auditing — Audit procedures — part 1: Auditing of environmental management systems

ISO 14012:199X, Guidelines for environmental auditing — Qualification criteria for environmental auditors

ISO 14050:199X, Environmental management — Vocabulary

3 Definitions

For the purpose of this International Standard, the following definitions apply.

audit conclusion: Professional judgement or opinion expressed by an auditor about the subject matter of the audit, based on and limited to reasoning the auditor has applied to audit findings.

audit criteria: Policies, practices, procedures, or requirements against which the auditor compares collected evidence about the subject matter.

Note 1: Criteria may include but are not limited to standards, guidelines, objectives, specified organizational requirements, and legislated or regulatory requirements.

audit findings: Result of the evaluation of the collected audit evidence compared against the agreed audit criteria.

Note 2: The findings provide the basis for the audit report.

audit team: Group of auditors, or a single auditor, designated to perform a given audit. The audit team may also include technical experts and auditors in training. One of the auditors on the audit team performs the function of lead auditor.

auditee: Organization to be audited.

auditor: Individual performing, an environmental audit, or part thereof, who meets the criteria specified in ISO 14012.

client: Organization commissioning the audit. The client may be the auditee, or any other organization that has the regulatory or contractual right to commission an audit.

environmental aspects: Components of an organization's activities, products, and services that are likely to interact with the environment.

environmental audit: Systematic, documented verification process of objectively obtaining and evaluating evidence to determine whether specified environmental activities, events, conditions, management systems, or information about these matters conform with audit criteria, and communicating the results of this process to the client.

APPENDIX B

environmental management system: Organizational structure, responsibilities, practices, procedures, processes, and resources for implementing and maintaining environmental management.

evidence: Verifiable information, records, or statements of fact.

Note 3: The evidence, which may be qualitative or quantitative, that is used by the auditor to determine whether audit criteria are met.

Note 4: Evidence is typically based on interviews, examination of documents, observation of activities and conditions, results of measurements, tests, or other means within the scope of the audit.

lead auditor: Auditor leading a specific environmental audit, who meets the criteria specified in ISO 14012.

Note 5: One of the responsibilities of the lead auditor is to supervise all members of the audit team.

organization: Company, corporation, government agency, firm, enterprise, institution, or association, or part thereof, whether incorporated or not, public or private, that has its own functions and administration.

subject matter: Specified environmental activities, events, conditions, management systems, or information about these matters.

technical expert: Individual who provides specific knowledge or expertise to the audit team, but who does not participate as an auditor.

verification: Process of authenticating evidence.

4 Requirements for an Environmental Audit

An environmental audit should focus on clearly defined and documented subject matter. The party (or parties) responsible for this subject matter should also be clearly defined and documented.

The auditor should only undertake the audit if, after consultation with the client, it is the auditor's opinion that:

- There is sufficient or appropriate information about the subject matter of the audit;
- There are adequate resources to support the audit process; and
- There is adequate cooperation from the auditee.

5 General Principles

5.1 Objectives and Scope

The audit should be based on objectives defined by the client. The scope is determined by the auditor in consultation with the client to meet these objectives. The scope describes the extent and boundaries of the audit.

The objectives and scope should be communicated to the auditee prior to the audit.

5.2 Objectivity, Independence, and Competence

To ensure the objectivity of the audit process and its findings and any conclusions, the members of the audit team should be independent of the activities they audit. The auditor should be objective, and free from bias and conflict of interest throughout the process.

The use of an external or internal auditor is at the discretion of the client. An auditor chosen from within the organization should not be accountable to those directly responsible for the subject matter being audited.

The auditor should possess an appropriate combination of knowledge, skills, and experience to carry out audit responsibilities. An auditor should meet the qualification criteria given in ISO 14012.

5.3 Due Professional Care

In the execution of an environmental audit, the auditor should use the care, diligence, skill, and judgment expected of any auditor in similar circumstances.

The relationship between the auditor and the client should be one of confidentiality and discretion. The audit team should not disclose information or documents obtained during the audit, and the final report, to any third party, without the expressed approval of the client and, where appropriate, the approval of the auditee.

The auditor should follow procedures that provide for adequate quality assurance. The auditor should apply these audit standards consistently and should seek authoritative interpretations when necessary.

5.4 Systematic Procedures

The environmental audit should be performed in accordance with these general principles and the guidelines developed for the appropriate type of environmental audit as defined in the various parts of ISO 14011.

To enhance consistency and reliability, the environmental audit should be conducted according to documented and well-defined methodologies and systematic procedures. Different types of environmental audits may require different methodologies and procedures. However, for any type of environmental audit, the methodologies and procedures should be consistent. The procedures for one type of audit differ from those of another only where it is essential to the specific character of a given type of audit.

5.5 Audit Criteria, Evidence, and Findings

An early and essential step in an environmental audit should be the determination of criteria. These criteria at an appropriate level of detail should be agreed between the auditor and the client, and then communicated to the auditee.

APPENDIX B

The auditor should collect, analyze, interpret, and document appropriate information to be used as evidence in an examination and evaluation process to determine whether the criteria are met.

Evidence should be of such a quality and quantity that competent auditors working independently of each other would reach similar findings from evaluating the same evidence against the same criteria.

5.6 Reliability of Findings and Conclusions

The environmental auditing process is designed to provide the client and auditor with the desired level of confidence in the reliability of the audit findings and any audit conclusions regarding the correspondence between evidence and criteria.

The environmental auditor should consider, throughout the audit, the risk of reaching an incorrect finding and the risk of reaching an incorrect conclusion, and should take these risks into account in planning and executing the audit.

The environmental auditor should obtain sufficient evidence to ensure that significant individual findings or aggregates of small findings, which could affect the audit conclusions, are taken into account.

The evidence collected will be a sample of the information available. This creates an element of uncertainty that is inherent to all environmental audits. All users of the results of environmental audits should recognize this uncertainty.

Note 6: This is partly due to the fact that an environmental audit is conducted during a limited period of time and at a limited cost.

5.7 Reporting

The audit findings or a summary thereof should be communicated to the client in a written report. Unless specifically excluded by the client, the auditee should receive a copy of the audit report.

Audit-related information that may be included in audit reports includes, but is not limited to:

- The identification of the organization audited and of the client;
- The identification of the auditee's representatives participating in the audit;
- The identification of the audit team members;
- The audit period;
- The agreed objectives and scope of the audit;
- The agreed criteria against which the audit was conducted;
- A summary of the audit process, including any obstacles encountered; and
- The audit conclusions.

The auditor in consultation with the client should determine which of these items, together with any additional items, should be included in the report.

Note 7: Normally, it should be the responsibility of the client or the auditee to determine any corrective action needed to respond to the findings. However, the auditor may include recommendations or opinions even there has been a prior agreement to do so with the client.

APPENDIX B

2nd Committee Draft   1 February 1995

ISO/CD 14011/1.2 "GUIDELINES FOR ENVIRONMENTAL AUDITING — AUDIT PROCEDURES — PART 1: AUDITING OF ENVIRONMENTAL MANAGEMENT SYSTEMS"

0 Introduction

Organizations of all kinds have a need to demonstrate environmental responsibility. The concept of Environmental Management Systems (EMS), with the associated practice of Environmental Auditing, has been advanced as one way to satisfy this need. These systems are intended to help an organization establish and continue to meet its environmental policies, objectives, standards, and other requirements.

This part of ISO 14011 provides procedures for the conduct of EMS audits. It is applicable to all types and sizes of organizations operating an EMS.

1 Scope

This part of ISO 14011 establishes audit procedures that provide for the planning and performance of an audit of an EMS to determine conformance with EMS audit criteria.

2 Normative References

The following standards contain provisions that, through reference in this text, constitute provisions of this part of ISO 14011. At the time of publication, the editions indicated were valid. All standards are subject to revision, and parties to agreements based on this part of ISO 14011 are encouraged to investigate the possibility of applying the most recent editions of the standards indicated below. Members of IEC and ISO maintain registers of currently valid International Standards.

ISO 14001:199X   Environmental Management Systems — Specification with guidance for use
ISO 14010:199X   Guidelines for environmental auditing — General principles of environmental auditing
ISO 14012:199X   Guidelines for environmental auditing — Qualification criteria for environmental auditors

3 Definitions

For the purposes of this part of ISO 14011 the definitions given in ISO 14010 and ISO 14001 apply together with the following definitions.

environmental management system audit: A systematic and documented verification process to objectively obtain and evaluate evidence to determine whether an organization's Environmental Management System conforms to the EMS audit criteria, and to communicate the results of this process to the client.

environmental management system audit criteria: Requirements derived from policies, practices, procedures, and other elements, as covered by ISO 14001 and, if applicable, any additional EMS requirements against which the auditor compares collected evidence about the subject matter.

4 Environmental Management System Audit Objectives, Roles, and Responsibilities

4.1 Audit Objectives

An EMS audit should have defined objectives; examples of typical objectives are as follows:

a) to determine conformance of an auditee's EMS against the EMS audit criteria;
b) to determine whether the auditee's EMS has been properly implemented and maintained;
c) to identify areas of potential improvement in the auditee's EMS;
d) to assess the ability of the internal management review process to ensure the continuing suitability and effectiveness of the EMS; and
e) to evaluate the EMS of an organization where there is a desire to establish a contractual relationship, such as with a potential supplier or a joint venture partner.

4.2 Roles, Responsibilities, and Activities

4.2.1 Lead Auditor

The lead auditor is responsible for ensuring the efficient and effective conduct and completion of the audit within the scope and audit plan approved by the client.

In addition, the lead auditor's responsibilities and activities should cover:

a) consulting with the client in determining the scope of the audit;
b) directing the activities of the audit team in accordance with the guidelines of ISO 14010 and this part of ISO 14011;
c) obtaining relevant background information necessary to meet the objectives of the audit, such as details of the auditee's activities, products, services, site, and immediate surroundings and details of previous audits;

d) forming the audit team giving consideration to potential conflicts of interest, and agreeing on its composition with the client;
e) preparing the audit plan with appropriate consultation with the client, auditee, and audit team members;
f) communicating the final audit plan to the audit team, auditee and the client;
g) coordinating the preparation of working documents and detailed procedures and briefing the audit team;
h) seeking to resolve any problems that arise during the audit;
i) recognizing when audit objectives become unattainable and reporting the reasons to the client and the auditee;
j) representing the audit team in discussions with the auditee, prior to, during, and after the audit;
k) notifying to the auditee observations of critical nonconformities without delay;
l) reporting on the audit clearly and conclusively within the time agreed with the client; and
m) making recommendations to the auditee for improvements to the EMS, if agreed in the scope of the audit.

4.2.2 Auditor

The auditor's responsibilities and activities should cover:

a) following the directions of and supporting the lead auditor;
b) planning and carrying out the assigned task objectively, effectively, and efficiently within the scope of the audit;
c) collecting and analyzing relevant and sufficient evidence to allow findings to be made and conclusions to be drawn regarding the audited EMS;
d) preparing working documents under the direction of the lead auditor;
e) documenting individual audit findings;
f) safeguarding documents pertaining to the audit and returning such documents as required; and
g) assisting in writing the audit report.

4.2.3 Audit Team

In selecting the audit team members, consideration should be given to:

a) qualifications as identified in ISO 14012;
b) the type of organization, activity, or function being audited;
c) the number, language skills, and expertise of the auditors;
d) any potential conflict of interest between the auditor and the auditee.

The audit team may also contain technical experts and auditors in training that are acceptable to the client, auditee, and lead auditor.

4.2.4 Client

The client's responsibilities and activities should cover:

a) determining the need for the audit;
b) contacting the auditee to obtain its full cooperation and initiate the process;
c) defining the objectives of the audit;
d) selecting the lead auditor or auditing organization and, if appropriate, approving the composition of the audit team;
e) providing appropriate authority and resources to conduct the audit;
f) consulting with the lead auditor to determine the scope of the audit;
g) approving the EMS audit criteria;
h) approving the audit plan; and
i) receiving the audit report and determining its distribution.

4.2.5 Auditee

The responsibilities and activities of the auditee should cover:

a) informing employees about the objectives and scope of the audit as necessary;
b) providing the facilities needed for the audit team in order to ensure an effective and efficient audit process;
c) appointing responsible and competent staff to accompany members of the audit team, to act as guides to the site and to ensure that auditors are aware of health, safety, and other appropriate requirements;
d) providing access to the facilities and personnel and relevant evidential material as requested by the auditors; and
e) cooperating with the auditors to permit the audit objectives to be achieved;
f) receiving a copy of the audit report unless specifically excluded by the client.

5 Auditing

5.1 Initiating the Audit

5.1.1 Audit Scope

The scope of the audit is determined by the client and the lead auditor. The scope describes the extent and boundaries of the audit in terms of factors such as physical location and organizational activities as well as the manner of reporting the audit results. The auditee should normally be consulted when determining the scope of the audit. Any subsequent changes to the audit scope need the agreement between the client and the lead auditor.

The resources committed to the audit should be sufficient to meet its intended scope.

5.1.2 Preliminary Document Review

At the beginning of the audit process, the lead auditor should review the organization's documentation such as environmental policy statements, programs, records, or manuals for meeting its EMS requirements. In doing so, use should be made of all appropriate background information on the auditee's organization. If the documentation is judged to be inadequate to carry out the audit the client should be informed. Further resources should not be expended until further instructions have been received from the client.

5.2 Preparing the Audit

5.2.1 Audit Plan

The audit plan should be designed to be flexible in order to permit changes in emphasis based on information gathered during the audit, and to permit effective use of resources.

The plan should include, if applicable:

a) the dates and places where the audit is be conducted;
b) the audit objectives and scope;
c) the audit criteria;
d) the procedures for auditing the auditee's EMS elements as appropriate for the auditee's organization;
e) identification of the functions and/or individuals within the auditee's organization having significant direct responsibilities regarding the subject matter of the audit;
f) identification of high-priority aspects of the auditee's EMS or activities;
g) identification of audit team members;
h) the working and reporting languages of the audit;
i) identification of reference documents;
j) identification of the auditee's organizational and functional units to be audited;
k) the expected time and duration for major audit activities;
l) the schedule of meetings to be held with the auditee's management;
m) confidentiality requirements;

n) report format and structure, expected date of issue and distribution of the audit report; and
o) document retention requirements.

The audit plan should be communicated to the client, the auditors, and the auditee. The client should review and approve the plan.

If the auditee objects to any provisions in the audit plan, such objections should be made known to the lead auditor. They should be resolved between the lead auditor, the auditee and the client before executing the audit. Any revised audit plan should be agreed between the parties concerned before executing the audit.

5.2.2 Audit Team Assignments

As appropriate, each audit team member should be assigned specific EMS elements, functions, or activities to audit and be instructed on the audit procedure to follow. Such assignments should be made by the lead auditor in consultation with the audit team members concerned. During the audit, the lead auditor may make changes to the work assignments to ensure the optimal achievement of the audit objectives.

5.2.3 Working Documents

The working documents required to facilitate the auditor's investigations may include:

a) forms for documenting supporting evidence and audit findings;
b) procedures and checklists used for evaluating EMS elements; and
c) records of meetings.

Working documents should be filed until audit completion; those involving confidential or proprietary information should be suitably safeguarded by the audit team members.

5.3 Executing the Audit

5.3.1 Opening Meeting

There should be an opening meeting. The purpose of an opening meeting is to:

a) introduce the members of the audit team to the auditee's management;
b) review the scope, objectives, and audit plan and agree on an audit timetable;

c) provide a short summary of the methods and procedures to be used to conduct the audit;
d) establish the official communication links between the audit team and the auditee;
e) confirm that the resources and facilities needed by the audit team are available;
f) confirm the time and date for the closing meeting;
g) promote the active participation by the auditee; and
h) review site safety and emergency procedures for the audit team.

5.3.2 Collecting Evidence

Evidence should be collected through interviews, examination of documents, and observation of activities and conditions. Indications of nonconformity to the EMS audit criteria should be recorded.

Information gathered through interviews should be verified by acquiring supporting information from independent sources, such as observations, records, and results of existing measurements. Nonverifiable statements should be recorded as such.

Sufficient objective evidence should be collected to verify that the auditee's EMS conforms to the EMS audit criteria.

Auditors should examine the basis of relevant sampling programs and the procedures for ensuring effective quality control of sampling and measurement processes.

5.3.3 Audit Findings

All significant audit findings should be recorded. All significant nonconformities should be documented.

The audit team should review all of their audit findings to determine where the EMS does not conform to the EMS audit criteria. The audit team should then ensure that these are documented in a clear, concise manner and supported by evidence.

Findings should be reviewed with the responsible auditee manager with a view to obtaining acknowledgement of all findings of nonconformity.

Note: If within the agreed scope, details of findings of conformity may also be documented, but with due care to avoid any implication of absolute assurance.

5.3.4 Closing Meeting with the Auditee

After completion of the evidence collection phase and prior to preparing an audit report, the auditors should hold a meeting with the auditee's management and those responsible for the functions audited. The main purpose of this meeting is to present audit findings to the auditee in such a manner as to ensure that they clearly understand and acknowledge the factual basis of the findings.

Disagreements should be resolved, if possible before the lead auditor issues the report. Final decisions on the significance and description of the findings ultimately rest with the lead auditor, though the auditee or client still may disagree with these findings.

5.4 Audit Reports and Records

5.4.1 Audit Report Preparation

The audit report is prepared under the direction of the lead auditor, who is responsible for its accuracy and completeness. The topics to be addressed in the audit report should be determined in consultation with the client.

5.4.2 Report Content

The audit report should be dated and signed by the lead auditor. The audit report should contain the audit findings or a summary thereof with reference to supporting evidence. Subject to agreement between the lead auditor and client the audit report should also include the following:

a) identification of the organization audited and of the client;
b) the auditee's representatives participating in the audit;
c) the audit team members;
d) the audit period;
e) the scope, objectives, and plan of the audit;
f) the agreed criteria, including a list of reference documents against which the audit is conducted;
g) the distribution list for the audit report; and
h) a statement of the confidential nature of the contents.

Subject to further agreement between the lead auditor and client the audit report may also include:

i) a summary of the audit process, including any obstacles encountered; and
j) audit conclusions such as:
  - EMS conformance to the EMS audit criteria;
  - whether the system is properly implemented and maintained; and
  - whether the internal review process is able to ensure the continuing suitability and effectiveness of the EMS.

5.4.3 Report Distribution

The audit report should be sent to the client by the lead auditor. Distribution of the audit report should be determined by the client in accordance with the audit plan. The auditee should receive a copy of the audit report unless specifically excluded by the client. Additional distribution of the report outside the auditee's organization requires the auditee's permission. Audit reports are the sole property of the client and confidentiality should be respected and appropriately safeguarded by the auditors and all report recipients.

The audit report should be issued within the agreed time period in accordance with the audit plan. If this is not possible, the reasons for the delay should be formally communicated to both the client and the auditee and a revised issue date established.

5.4.4 Record Retention

All documents pertaining to the audit should be retained by agreement between the client, the lead auditor, and the auditee, and in accordance with any applicable requirements. Auditors may not disclose any documents without the express permission of the client and the auditee.

6 Audit Completion

The audit is completed once all activities relating to the agreement between the client, the auditee, and the lead auditor have been concluded.

2nd Committee Draft   1 February 1995

ISO/CD 14012.2 "GUIDELINES FOR ENVIRONMENTAL AUDITING — QUALIFICATION CRITERIA FOR ENVIRONMENTAL AUDITORS"

0 Introduction

To support the application of environmental management systems and environmental auditing, guidance is needed on qualification criteria for environmental auditors. This International Standard aims to provide such guidance.

It addresses the qualification criteria for auditors and lead auditors. Criteria for the collective qualifications of audit teams are not included; reference should be made to ISO 14011 for further information on that subject.

This International Standard is applicable to both internal and external auditors.

Internal auditors need the same set of competencies as external auditors, but might not meet in all respects the detailed criteria described herein, depending upon such factors as:

- The size, nature, complexity, and environmental impacts of the organization; and
- The rate of development of the relevant expertise and experience within the organization.

1 Scope

This International Standard provides guidance on qualification criteria for environmental auditors. It is applicable to the selection of auditors to perform environmental audits as described in various parts of ISO 14011.

2 Normative References

The following standards contain provisions that, through reference in this text, constitute provisions of this International Standard. At the time of publication, the editions indicated were valid. All standards are subject to revision, and parties to agreements based on this International Standard are encouraged to investigate the possibility of applying the most recent editions of the standard indicated below. Members of IEC and ISO maintain registers of currently valid International Standards.

ISO 14001:199X   Environmental management systems —Specification and guidance for use

ISO 14010:199X   Guidelines for environmental auditing — General principles of environmental auditing

ISO 14011-1:199X Guidelines for environmental auditing — Audit procedures — part 1: Auditing of environmental management systems
ISO 14050:199X Environmental management — Vocabulary

3 Definitions

For the purposes of this International Standard, the definitions given in ISO 14010 and ISO 14050 apply, together with the following:

appropriate work experience: Work experience that contributes to the development of skills and understanding in some or all of the following:

- Environmental science and technology;
- Technical and environmental aspects of facility operations;
- Relevant requirements of environmental laws, regulations, and related documents;
- Environmental management systems and standards; and
- Audit procedures, processes, and techniques.

auditor (environmental): An individual performing an audit, or part thereof, who meets the criteria specified in this International Standard.

lead auditor (environmental): An auditor leading a specific audit, who meets the criteria specified in this International Standard.

degree: A recognized national or international degree, or equivalent qualification, normally obtained, after secondary education, through a minimum of 3 years formal full-time, or equivalent part-time study.

secondary education: That part of the national educational system that comes after the primary or elementary stage, but that is completed immediately prior to entrance to a university or similar establishment.

4 Education and Work Experience

Auditors should have completed at least secondary education, or equivalent.

Auditors who have completed secondary education or equivalent only should have a minimum of 5 years of appropriate work experience. This criterion may be reduced by satisfactory completion of formal full-time or part-time education, the contents of which address some or all of the topics listed in clause 3. 1. Any reduction should not exceed the total period of such education addressing those topics, and the total reduction should not exceed 1 year.

Auditors who have obtained a degree should have a minimum of 4 years appropriate work experience. This criterion may be reduced by satisfactory completion of formal full-time or part-time education, the contents of which address the topics listed in clause 3.1. Any reduction should not exceed the total period of such education addressing those topics, and the total reduction should not exceed 2 years.

5 Auditor Training

In addition to the criteria described above, auditors should have completed both formal training and on-the-job training, to develop competence in carrying out environmental audits. Such training may be provided by the auditor's own organization, or by an external organization.

Competence achieved through training should be demonstrated by suitable means, examples of which are provided in Annex A.

5.1 Formal Training

Formal training should address:

- Environmental science and technology;
- Technical and environmental aspects of facility operations;
- Relevant requirements of environmental laws, regulations, and related documents;
- Environmental management systems and standards against which audits may be performed; and
- Audit procedures, processes, and techniques.

The criterion for formal training in some or all of these areas may be waived if competence can be demonstrated through accredited examinations or relevant professional qualifications.

5.2 On-the-Job Training

An auditor should have completed a period of on-the-job training for a total of 20 equivalent workdays of auditing, and for a minimum of four audits. This should include involvement in the entire audit process under the supervision and guidance of the lead auditor. This on-the-job training should occur within a period of not more than 3 consecutive years.

6 Objective Evidence of Education, Experience, and Training

Individuals should keep objective evidence of their education, experience, and training.

7 Personal Attributes and Skills

Auditors should possess personal attributes and skills that include, but are not limited to:

- Competence in clearly and fluently expressing concepts and ideas, orally and in writing;
- Interpersonal skills conducive to the effective and efficient performance of the audit, such as diplomacy, tact, and the ability to listen;
- The ability to maintain independence and objectivity sufficient to permit the accomplishment of auditor responsibilities;
- Skills of personal organization necessary to the effective and efficient performance of the audit; and
- Ability to reach sound judgements based on objective evidence.

8 Lead Auditor

The lead auditor for an environmental audit should be an auditor who shows a thorough understanding and application of those personal attributes and skills necessary to ensure effective and efficient management and leadership of the audit process, and who meets the following additional criteria:
Either:

- Demonstration of meeting the above criteria to the audit program management, or others, by means such as interviews, observation, references, and quality assurance programs.

Or:

- Participation in the entire audit process for a total of 15 additional equivalent workdays of auditing, for a minimum of three additional complete audits; and
- Participation as acting lead auditor, under the supervision and guidance of a lead auditor, for at least one of the above three audits.

These additional criteria for lead auditor should be met within a period of not more than 3 consecutive years.

9 Maintenance of Competence

Auditors should maintain their competence by ensuring the currency of their knowledge of:

- Environmental management systems and related standards;
- Auditing processes, procedures, and techniques;
- Relevant environmental laws, regulations, and related documents;
- Aspects of relevant environmental science and technology; and
- Appropriate technical and environmental aspects of facility operations.

They should ensure that their experience in the execution of audits is current, and should participate in refresher training as necessary.

10 Due Professional Care

Auditors should exercise due professional care, as addressed in ISO 14010 and ISO 14011/1, and adhere to an appropriate code of ethics.

11 Language

No audit team members should participate unsupported in an audit where they are not able to communicate effectively in the language necessary for performing their responsibilities. When necessary, support should be obtained from a person with the necessary language skills, who is not subject to pressures that would affect the performance of the audit.

Annex A (Informative) — Evaluating the Qualifications of Environmental Auditors

A.1 General

This annex provides guidance for evaluating the qualifications of environmental auditors as defined in this International Standard.

A.2 Evaluation Process

This International Standard may be implemented by the establishment and operation of an evaluation process. The process may be internal or external to the auditor's audit program management. Its purpose is to evaluate the qualifications of environmental auditors.

This process should be directed by an individual or individuals having current understanding and experience of auditing operations.

The environmental auditor evaluation process may be subject to a quality assurance program.

A.3 Evaluations of Education, Work Experience, Training, and Personal Attributes

There should be evidence to show that environmental auditors have acquired and maintained the necessary education, work experience, training, and personal attributes as described in this International Standard. The evaluation process should include some of the following methods:

- Interviews with candidates;
- Written and/or oral assessment or other suitable means;

- Review of candidates' written work;
- Discussions with former employers, colleagues, etc.;
- Role playing;
- Peer observation under actual audit conditions;
- Reviewing records of education, experience, and training as defined in this International Standard;
- Consideration of professional certifications and qualifications.

Annex B (Informative) — Environmental Auditor Registration Body

B.1 General

This annex contains guidance on the development of a body to ensure a consistent approach to the registration of environmental auditors.

B.2 Auditor Registration

If it is appropriate to establish a body for ensuring that environmental auditors are registered in a consistent manner, such a body should be independent and the following guidelines should apply.

The body may act to register environmental auditors directly or accredit other organizations who in turn register environmental auditors to the criteria contained in this International Standard.

The body should establish an evaluation process consistent with that contained in annex A of this International Standard. The process should be subject to a quality assurance program.

The body should keep a register of environmental auditors who currently meet the criteria specified in this International Standard.

Appendix C
ORGANIZATIONAL STRUCTURE OF ISO 14000/TC 207:
Environmental Management and United States Participation Chart

ENVIRONMENTAL AND SAFETY AUDITING

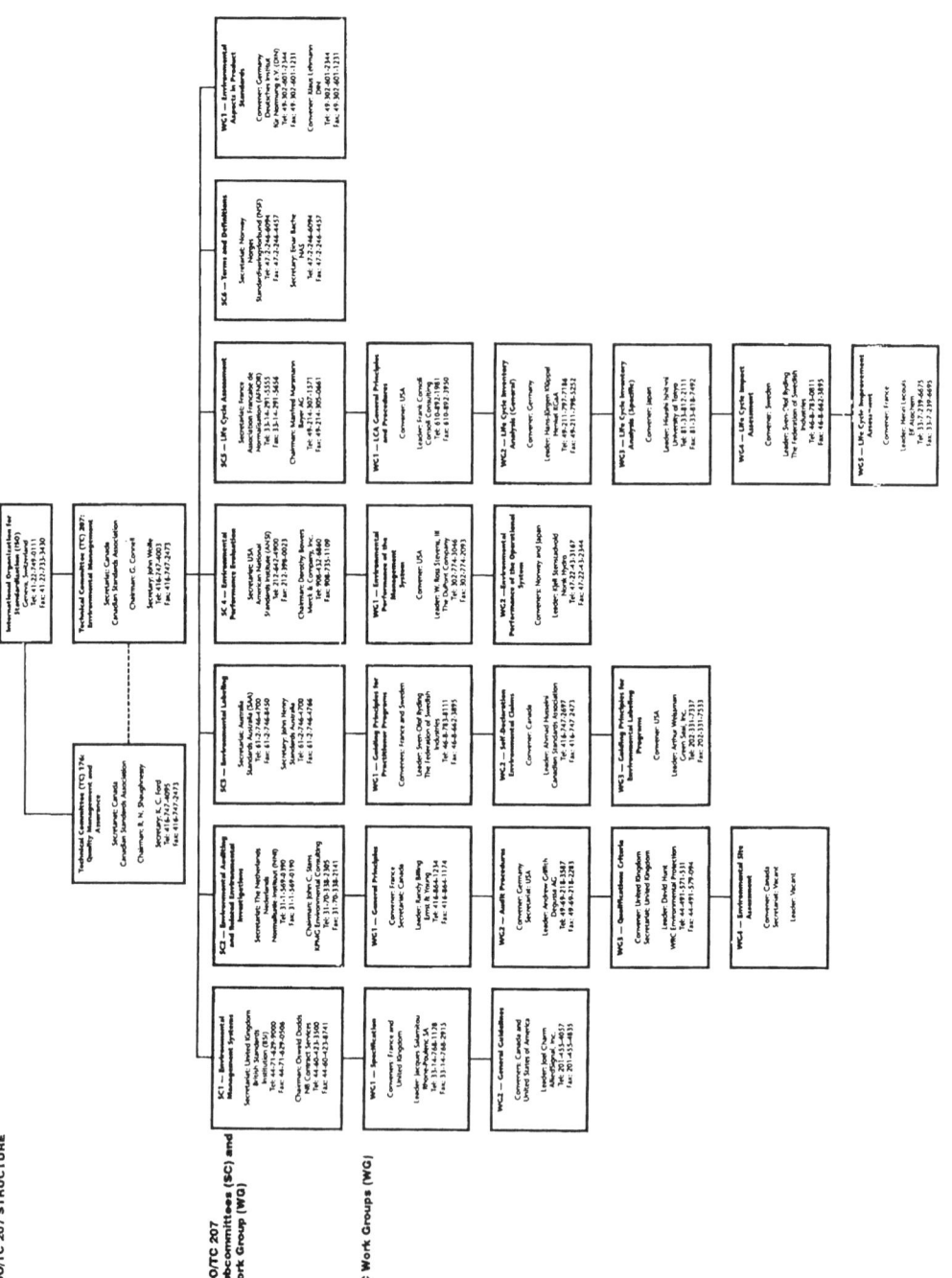

APPENDIX C

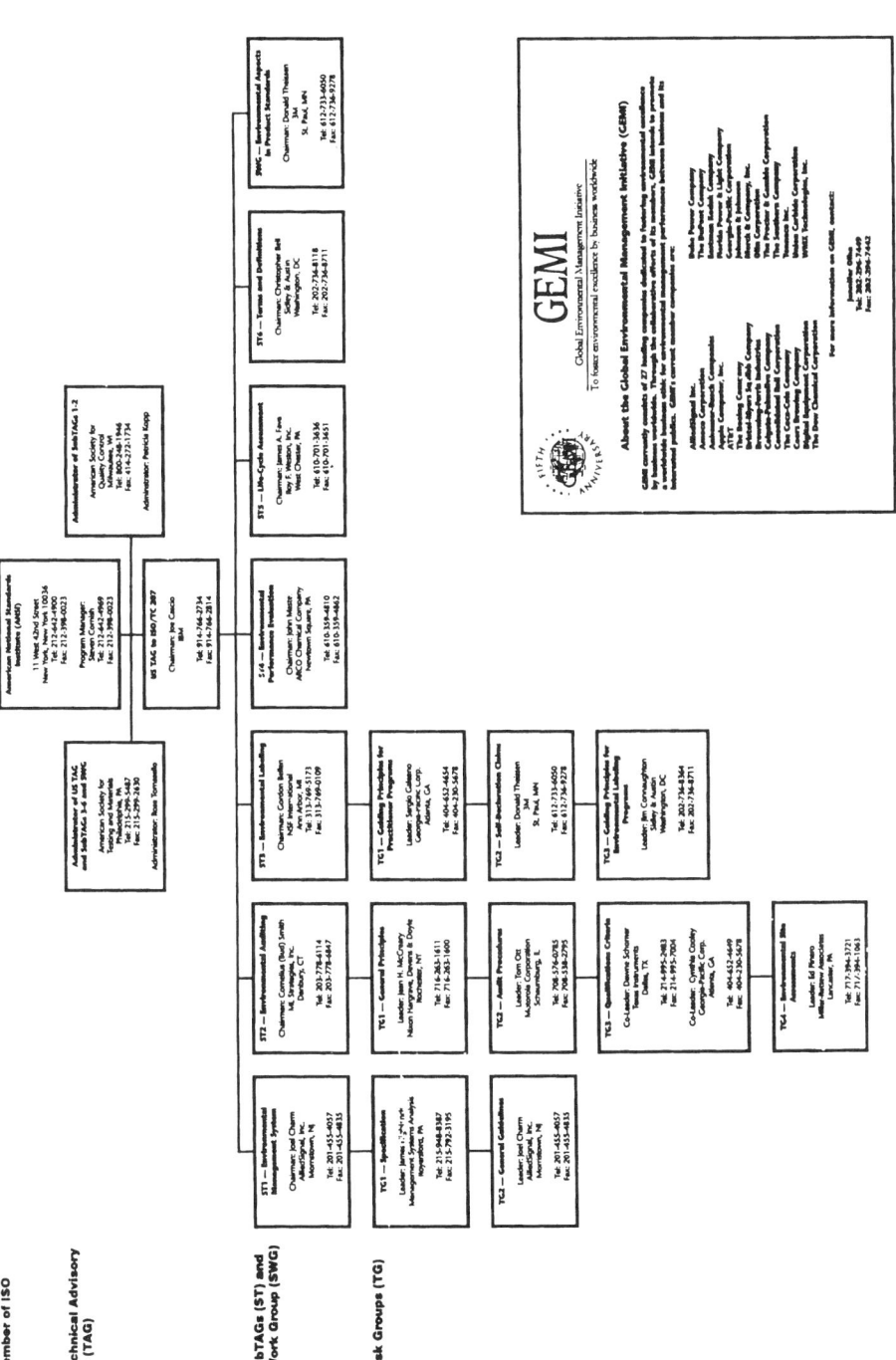

Organizational structure of ISO 14000/TC 207: Environmental management and United States participation.

Appendix D
GLOSSARY

Definitions 219

Acronyms 221

DEFINITIONS

Contingent Liability — A contingent liability, in an environmental context, is a liability where the outcome is uncertain and the cost or dollar amount that the liability represents is not fixed.

Due Diligence — Due diligence is a self-motivated, conscientious, systematic effort to meet regulatory criteria.

Economic Benefit Component — The major part of a penalty that represents the economic advantage a violator gains through its noncompliance. The minor part of the penalty is the gravity component.

Environmental Audit — Environmental auditing is a systematic, documented, periodic, and objective review by regulated entities of facility operations and practices related to meeting environmental requirements.

Grace Period — A period of 90 days to correct any violation without any penalty being levied.

Gravity Component — That minor portion of a penalty that represents the seriousness or punitive portion. The major part of the penalty is the economic benefit component.

Green Marketing — The selling of a company's environmental philosophy, policy, and programs.

Green Report — An annual report that summarizes the environmental programs, progress, and accomplishments of a company.

Regulated Entities — Regulated entities include private firms and public agencies with facilities subject to environmental regulations. Public agencies can include federal, state, and local agencies.

Safe Harbor — The establishment through regulation of legal protective measures and privileges.

Superfund — The Comprehensive Environmental Response Compensation and Liability Act (CERCLA), so-called because of its funding mechanism focusing on industry taxes that amassed large amounts of money.

Sustainable Development — The meeting of the needs of the present without compromising the ability of future generations to meet their own needs. Sustainable Development relies on the sustainable use of natural resources.

Sustainable Development Programs — The integrated management of company resources, environmental resources, finances, regulatory compliance, and public image for extended company growth and development.

ACRONYMS

<	less than
>	greater than
ANSI	American National Standards Institute
AIHA	American Industrial Hygiene Association
API	American Petroleum Institute
BS7750	British Standard 7750
CAA	Clean Air Act
CAAA	Clean Air Act Amendments
CD	Committee Draft (ISO term)
CEEC	Corporate Environmental Enforcement Council
CERCLA	Comprehensive Environmental Response Compensation and Liability Act
CERE	Coalition for Environmentally Responsible Economics
CEQ	Council on Environmental Quality
CFR	Code of Federal Regulations
CIH	Certified Industrial Hygienist
CMA	Chemical Manufacturers Association
CMPG	Compliance Management Policy Group
CWA	Clean Water Act
DIS	Draft International Standard (ISO term)
DOJ	United States Department of Justice
EA	Environmental Assessment
EAR	Environmental Auditing Roundtable
EARA	Environmental Auditing Registration Association
EEI	Energy, Environment and Industrial Hygiene Corporation
EIS	Environmental Impact Statement
EITF	Emerging Issues Task Force

ELP	Environmental Leadership Program
EMS	Environmental Management System
EPA	United States Environmental Protection Agency
EPCRA	Emergency Planning and Community Right-to-Know Act
ES&H	Environmental, Safety and Health
ESAP	Environmental Self-Assessment Program
FASB	Financial Accounting Standard Board
FEDPLAN	Federal Agency Environmental Management Plan
FFCA	Federal Facilities Compliance Act
FFEO	Federal Facility Enforcement Office
FOIA	Freedom of Information Act
FR	*Federal Register*
gal	gallons
GAO	Government Accounting Office
GEMI	Global Environmental Management Initiative
HCRS	Heritage Conservation and Recreation Service
HR	House of Representatives
HSWA	Hazardous and Solid Waste Amendments
ICC	International Chamber of Commerce
ISO	International Standards Organization
kg	kilogram
L.L.P.	Limited Liability Partnership
MD&A	Management Discussions and Analysis
mo.	month
NAAQS	National Ambient Air Quality Standards
NEPA	National Environmental Policy Act
NPDES	National Pollutant Discharge Elimination System
OECA	Office of Enforcement and Compliance Assurance
OMB	Office of Management and Budget
OSH Act	Occupational Safety and Health Act

OSHA	Occupational Safety and Health Administration
OHSMS	Occupational Health and Safety Management System
POTW	Publicly Owned Treatment Work
PPA	Pollution Prevention Act
RCRA	Resource Conservation and Recovery Act
Ref.	Reference
RI/FS	Remedial Investigation/Feasibility Study
SARA	Superfund Amendments Reauthorization Act
SDWA	Safe Drinking Water Act
SEC	Securities and Exchange Commission
SIP	State Implementation Plans
SWDA	Solid Waste Disposal Act
t	ton(s)
TC	Technical Committee 207 of the ISO
TSCA	Toxic Substances Control Act
TSD	treatment, storage, or disposal
U.S.C.	United States Code
UST	Underground Storage Tanks
VOC	Volatile Organic Compound
VPP	Voluntary Protection Program (OSHA Program)
yr	year

INDEX

A

American Petroleum Institute, 60
Arcadian Case History, 75
Attorney-Client Privilege, 6, 12-13, 61, 66-67,
　　69, 84, 87, 89, 95-96, Appendix A
　Case Histories, 71-72
Audit Programs, 83-101
　Audit Privilege, 17-18, 30, 31, 61, 70, 77,
　　157, 166, 181, 186
　Baseline Opportunity, 2, 88-90
　Compliance Audit, 21-23, 25, 29, 36, 48,
　　52, 65, 68, 89, 100, 104, 112, 164-
　　165, 176, 184
　Controversies, 5-7
　Cost, 12
　Definitions, 1-4
　Educational Process, 10-11
　Policies, 6, 18-20, 20-21, 26-32,
　　Appendix A
　Program, 19, 83-101
　Risk/Benefit, 1, 2, 5, 9-14, 16-18, 33, 84,
　　Appendix A, Appendix B
　Voluntary/Mandatory, 5-6, 13, 15, 29, 45,
　　49, 56, 59-61, 72, 76, 111,
　　Appendix A

B

British Standard 7750, 23, 48, 112, 113-115

C

Case Histories, 71-72
　Arcadian Case History, 75
　Caterpillar Case History, 75
　Hartford Roofing Case History, 75
　Olen Properties vs. Sheldahl, Inc., 71
　Reichhold Chemicals, Inc. vs. Textron,
　　Inc., 72
　United States vs. Chevron U.S.A. Inc., 71
　United States vs. Dexter Corp., 72
Caterpillar Case History, 75
Coalition of Environmentally Responsible
　　Economies, 51-52
Chemical Manufacturers' Association, 7,
　　49-50
Clean Air Act, 104
Clean Water Act, 15, 72, 104
Common Sense Initiative, 15, 22, 36
Compliance Management Policy Group, 60
Corporate Environmental Enforcement
　　Council, 59
Correction and Remediation, 13, 21, 29-30, 46,
　　61, 82, 88, 166, 171-172, 178, 185
Criminal Violation, 16-18, 27, 35

D

Demonstration Program, 76
Department of Justice, 5, 15, 16-18, 27, 32,
　　69, 70, 74, 165
　Criminal Cases, 71-72
　House of Representatives Bill 1047 - See
　　'Voluntary Environmental Self-
　　Evaluation'
　Sentencing Guideline, 16, 73
Disclosure, 2, 5-6, 11, 12, 13, 15, 16, 26, 29,
　　45, 65-71, 121-130, Appendix A
　Federal Agencies, 15, 16, 26, 69-70
　Grace Period, 23, 45, 46
　Occupational Safety and Health
　　Administration, 81-82
　Securities and Exchange Commission, 2,
　　11, 15, 121-130
　Policies, 26, 29, Appendix A

E

Environmental Auditing Round Table, 7, 56, 132
Economic Benefits, 27, 32, 62, 153, 183, 186
Enforcement Statistics, 72
Environmental Auditors Registration Association, 48, 119
Environmental Crimes Act, 15
Environmental Management System, 9, 21-23, 47-48, 56, 98, 100-101, 111-117, Appendix B
  Audits, 22
  Environmental Leadership Program Pilot Projects, 23-25
  International Standards Organization, 9, 21, 47-48, 100-101, 111-117
  Root Cause, 2, 4, 95, 96-98, 100, 164
Environmental Protection Agency, 18-32, Appendix A
  Environmental Leadership Program, 21-26
  Interim Policy, 20-21
  Office of Enforcement and Compliance Assurance, 18, 157
  Environmental Auditing Policy Statement of 1986, 18-20, Appendix A
  Self-Policing Policy of 1995, 26-32, Appendix A
  Preemption Over Other Enforcement Agencies, 18-19
Environmental Secrecy Act, 32
Environmental Self-Assessment Program, 52-56, 98, 100
European Community Eco-Management, 48, 103
Evidentiary Privilege, 17, 18, 32, 180-181
Executive Orders, 110

F

Federal Facility, 32-43
  Agency Mission, 38
  Audits, 32, 37-42
  Contractor Facility Arrangement, 41
  Federal Agency Environmental Management Plan, 40
  Federal Budget Cycle, 40
  Federal Facilities Compliance Act, 35
  Federal Facility Enforcement Office, 15, 36-37
  Freedom of Information Act, 42
  National Security, 39

G

Government Accounting Office, 33-34
Global Environmental Management Initiative, 7, 52
Gravity-Based Penalty, 13, 27, 70, 170, 172, 173-174, 183
Green Marketing, 9, 11, 101, 112, 219
Green Reports, 2, 11, 101, 110, 219

H

Hartford Roofing Case History, 75
House of Representatives Bill 1047 - See 'Voluntary Environmental Self-Evaluation Act'

I

Immunity, 15, 16-18, 31, 32, 35, 62, 70, 178, 186
  Department of Justice, 16-18
  Sovereign 15, 35
International Chamber of Commerce, 50-56,
International Standards Organization, 9, 21, 47-48, 100-101, 111-117 Appendix B

L

Liability Assurance, 10

M

Maine 200 Program, 79
Merit Program, 77

O

Occupational Safety and Health Act, 103
Occupational Safety and Health Administration, 74
  Auditing Policy, 81-82
  Case Histories, 75
  Egregious Penalty Policy, 75
  Enforcement, 74-76
  Fines, 75
  Maine 200 Program, 79
  Reform, 79-81
  State Safety Programs, 79-78
  Voluntary Protection Program, 22, 76-78
Olen Properties vs. Sheldahl, Inc., 71

P

Pollution Prevention, 15, 21, 22, 25, 26, 35, 36, 37, 48, 92, 99, 106, 110, 112, 113
Pollution Prevention Act, 106
Price Waterhouse Survey, 1, 31, 45-47, 84, 88, 95, 122, 126, 180
Project Xl, 22

R

Recurrence or Repeat Violation, 27, 28, 30, 32, 171, 177, 178, 179, 185
Reichhold Chemicals, Inc. vs. Textron, Inc., 72

S

Safe Harbor, 27, 28, 31, 45, 128
Safeguards, 30, 171
Safety, 4, 17, 22, 25, 74-76, 92, 93, 94, 99, 103
   Auditor's Personal Safety, 135
   Enforcement Case Histories, 75
   International Standards Organization, 112, 117
   Occupational Safety and Health Act, 103
   Public Safety, 17
   State Safety Programs, 79-78
   Voluntary Protection Program, 76-78
Superfund Amendments Reauthorization Act, 15, 105
Safe Drinking Water Act, 15, 107
Securities Exchange Commission, 11, 15, 121-130
   Annual Report, 2, 122
   Disclosure, 2, 15, 125-126, 129
   Enforcement, 122-123
   Environmental Accounting, 126
   Environmental Liability Management, 11, 122-124
   Recording of Environmental Lability, 122-123
Self Evaluation Privilege, 6, 15, 59-60, 166
Self-Critical Analysis, 66, 68, 72
Senate Environmental Auditing Bill, 15
Solid Waste Disposal Act, 15, 106
Star Program, 77

T

Third-Party Audits, 15, 22, 12, 98, 101, 116, Appendix A

U

United States vs. Chevron U.S.A. Inc., 71
United States vs. Dexter Corp, 72

V

Voluntary Environmental Self-Evaluation Act, 16-18

W

World Bank Group, 7, 63, 132